T0146110

Creatures Born of Mud and Slime

Singleton Center Books in Premodern Europe

The Charles S. Singleton Center for the Study of Premodern Europe of the Johns Hopkins University is an interdisciplinary consortium of humanities scholars on the university's faculty. Established in 2008, the Singleton Center fosters research of the European world in the Late Classical, Medieval, Renaissance, and Early Modern periods. The Singleton Center sponsors graduate research abroad and faculty-led initiatives to partner with European institutions of higher learning, as well as educational activities, lectures and lecture series by prominent scholars, and many other scholarly activities on the Johns Hopkins campus in Baltimore and at European venues. The center is named after Charles S. Singleton (1909–85), the renowned scholar of medieval literature who taught for most of his career at the Johns Hopkins University.

Every two years the Singleton Center organizes the Singleton Distinguished Lecture Series, which invites a prominent scholar of premodern Europe to the Homewood campus of Johns Hopkins to present three lectures on a common theme. The series was inaugurated in October 2010 by Professor Nancy Siraisi on the subject of epistolary networks in Renaissance medicine. The present volume is based upon lectures by Daryn Lehoux delivered in October 2014.

Lawrence M. Principe
Director, Singleton Center

CREATURES BORN *of* MUD AND SLIME

The Wonder and Complexity of Spontaneous Generation

DARYN LEHOUX

Johns Hopkins University Press
Baltimore

This book was brought to publication with the generous assistance of the Singleton Center for the Study of Premodern Europe.

Printed in the United States of America on acid-free paper
9 8 7 6 5 4 3 2 1

Johns Hopkins University Press
2715 North Charles Street
Baltimore, Maryland 21218-4363
www.press.jhu.edu

Library of Congress Cataloging-in-Publication Data

Names: Lehoux, Daryn, 1968– author.
Title: Creatures born of mud and slime : the wonder and complexity of spontaneous generation / Daryn Lehoux.
Description: Baltimore : Johns Hopkins University Press, [2017] | Series: Singleton Center books in premodern Europe | Includes bibliographical references and index.
Identifiers: LCCN 2017007358| ISBN 9781421423814 (hardcover : alk. paper) | ISBN 1421423812 (hardcover : alk. paper) | ISBN 9781421423821 (electronic) | ISBN 1421423820 (electronic)
Subjects: LCSH: Spontaneous generation—History. | Evolution (Biology)—History.
Classification: LCC QH325 .L44 2017 | DDC 576.8/3—dc23
LC record available at https://lccn.loc.gov/2017007358

A catalog record for this book is available from the British Library.

Special discounts are available for bulk purchases of this book. For more information, please contact Special Sales at 410-516-6936 or specialsales@press.jhu.edu.

Johns Hopkins University Press uses environmentally friendly book materials, including recycled text paper that is composed of at least 30 percent post-consumer waste, whenever possible.

To the fond memory of my teacher and friend Marcellus Martyr. When I was a teenager skipping lunch to attend his eleventh-grade philosophy class, he put a copy of Plato's Parmenides *into my hands, and changed everything.*

CONTENTS

ACKNOWLEDGMENTS

I would like to warmly thank the Singleton Center for the Study of Premodern Europe for its very kind invitation to deliver the Distinguished Lecture Series in 2014, and Larry Principe, in particular, for his helpful comments on this material as it developed. For comments and discussion on individual chapters, I thank Gábor Betegh, Christopher Celenza, Lorraine Daston, Jay Foster, Darin Hayton, Devin Henry, Stephen Menn, Jon Miller, Matt McAdam, Josh Mozersky, Carla Nappi, Dan-el Padilla Peralta, James Pitts, Courtney Roby, Sergio Sismondo, and Robert Wardy. Thanks also to the anonymous readers from Johns Hopkins University Press, whose thoughtful comments clarified more than a few of the issues in my mind.

Various parts of the book benefited greatly from discussion with audiences at Johns Hopkins, Bielefeld, Bryn Mawr, Cambridge, Columbia, Cornell, Humboldt Universität, Manchester, Queen's, Villanova, and the University of Western Ontario. Laura Bevilacqua, Matthew Chandler, Grant Shrama, and Julianna Will provided invaluable research assistance. I would like to thank the Social Sciences and Humanities Research Council of Canada, the Singleton Center, and Queen's University for funding this research. Last but certainly not least, I would like to thank my family for their unending support.

Creatures Born of Mud and Slime

Introduction

SOMEWHERE IN THIS marvelous universe of ours, at least once in its 13.8-billion-year history, life came into being from nonliving matter through some kind of chemical process. We don't yet know where or when this happened, but the fact that we are here to think about it at all is a trivial—but quite definitive—proof that it did.

The other morning I opened the cupboard door where we keep a little bucket for food scraps (my city composts its organic garbage), and the cupboard fairly erupted with fruit flies that were not there the previous night. What did *not* happen in the night, I don't believe, was that those flies simply sprang to life, somehow generated directly from the nonliving matter rotting in my compost bucket. I am perfectly happy with the idea that a complex enough chemical soup in the hot, turbid waters of the early earth (or at least of some planet) could have—and indeed, must have—given birth to self-replicating molecules and eventually to living cells. But those, one wants to believe, must have been exceptional circumstances. I am not at all comfortable with the idea that some or all of the plants, animals, or insects we see around us every day could ever come into existence spontaneously (which is to say, without parents and without seed) and that their generation happens from nothing more than the putrefaction of nonliving matter.

That being said, it is a curious fact in the history of the sciences that this is precisely what we did believe for a considerably long time.[1] Indeed, if we think about the great dismissals of old ideas that speckle the period often referred to as the scientific revolution of the sixteenth and seventeenth centuries—Galileo punching against Aristotelian physics, Copernicus throwing Ptolemy's cosmology up in the air—what we do not find is a stable and wholesale dismissal of this idea that some kinds of living entities are quite regularly generated spontaneously out of other matter.[2] Looking at how the debates play out over the centuries, we find the cast of characters variously expanding and contracting: Aristotle included some—but not all—insects, as well as eels, snails, sponges, and some shellfish, but Avicenna went so far as to include, under certain circumstances, human beings (this was how the world was repopulated after the flood, for example). By the early modern period, the list had again shrunk back down—in size if not, perhaps, in number of species—to smaller and less "perfect" animals, and then finally, by the nineteenth century, it was confined to microorganisms alone. The fact that even toward the end of its life, spontaneous generation was responsible for the birth of billions of individuals and innumerable species (they may have been small, but there were a lot of them) may strike us as counterintuitive. Starting with a handful of nameable species in Aristotle and then watching that list expand is, perhaps, not what we should expect. This is not a simple story of progress from thinking that there were many animals that were spontaneously generated, to thinking there were a few, to thinking there were none.

Here we find reason to pause: that the idea of spontaneous generation survived careful experiment after careful experiment, radical thinker after radical thinker, from remotest antiquity right through to very recent modernity—it survives even past the invention of the internal combustion engine—is truly remarkable. I can't say that I can think of a single instance of what we would now call a wrong scientific idea from antiquity that has had a more recent end.[3] Spontaneous generation was, in that sense, perhaps the last stand of the ancient scientific worldview.

Why is that?

It is often said in discussions of spontaneous generation that the idea is a kind of "common-sense" way of offering a simple, perhaps even simplistic, explanation for the way in which maggots just seem to "spring up" from rotting meat,[4] or of accounting for the generation of difficult-to-observe animals such as eels, whose reproductive mysteries are complicated by vast migrations and whose young are exceedingly difficult to find or to recognize (we did not completely figure out eels until well into the twentieth century). This is perhaps true to a certain extent, but when the history of the sciences is told as the history of people bumbling around without testing their common-sense notions until the fateful day when some heroically unblinkered investigator finally puts those ideas to an actual empirical test—a Francesco Redi or a Louis Pasteur—we end up with a story that is as unfair as it is uninteresting. Instead, an important theme I want to develop in what follows is just how careful our investigators into spontaneous generation were, how willing they were to challenge old chestnuts and to try and find ways to test their ideas about generation, and how painstaking and, often enough, parsimonious they were about assigning animals to the category of spontaneous generation. Far from being just common sense, conservatively or unreflectively clung to, spontaneous generation was very carefully thought about, thoroughly and intelligently elaborated, and continually questioned and tested.

This is not to say that doubts about the very idea of spontaneous generation were floated on a regular basis—they were not—but that on a case-by-case basis it was asked again and again whether this insect, that mammal, these shellfish, were really generated spontaneously or whether it could be shown that they generated sexually instead. The result was a continual rethinking about which animals or plants were generated spontaneously, under what conditions, and by what mechanisms. Over time, individual animals, or even whole groups of them, come to be understood as generating sexually, but it takes a remarkably long time for the possibility of spontaneous generation to be jettisoned entirely. This is not, I argue, through carelessness or gullibility. It is because the evidence for spontaneous generation was so very compelling. In fact, it stands up to testing quite rigorously. It is precisely because of this strong evidential base that spontaneous generation had, as E. I. Mendelsohn aptly put it, so many lives:

> What is curious about spontaneous generation is that its overthrow as a paradigm has been celebrated at least three times: once each in the seventeenth, eighteenth and nineteenth centuries. The heroes of each of these scientific revolutions can be readily identified: Redi and Leeuwenhoek working primarily on insect life cycles in the seventeenth century; Spallanzani who was concerned with the Infusoria in the eighteenth century; and Pasteur who once again "disproved" spontaneous generation through his work on bacteria in the nineteenth . . . It is clear that spontaneous generation, as a paradigmatic concept, did not die each time, but rather in descending order of size of organism its usefulness as an explanatory model for the generation of organisms was replaced.[5]

That spontaneous generation stood on such a solid empirical footing for so long is one of the main themes of this book, and I will return to it throughout, asking again and again what observation, what experiment or experience was performed, sought out, or reported to justify claims not only about whether a particular animal was spontaneously generated but also about the very biological mechanism of spontaneous generation, which was often difficult for our authors to work out.

A second theme I want to develop is that of authority. Our first author in this study, Aristotle, had a tremendous influence on the Arabic and Latin Middle Ages, the Renaissance, and the early modern period alike.[6] Here I want to look at not just Aristotle's role as a later authority but also his own use of authorities—how do he and other authors handle reports they get from fishermen, from beekeepers, from earlier writers? I hope to show on this latter question that Aristotle and others are more meticulous than is often noticed, both about how they report their sources' findings to the reader, as well as how consistent they are (for better or for worse) about whom they trust, and why, and when. Testimony plays a large role in spontaneous generation debates, as we shall see, because testimony plays a large role in all scientific endeavour,[7] and also because many of the animals thought to be possibly or definitely spontaneously generated, or the conditions under which their generation happened, were commonly accessible only to people who worked and lived in certain places or under certain conditions (on the sea, in far-off lands, after a particularly bad drought). Some of the ways in which Aristotle handled his authorities, his witnesses, and the ways in which he held their claims up to his own considerable experiences to test their plausibility, are mirrored in the ways that later authors held up Aristotle's account to their own experiences, their own understandings, to test *their* plausibility. We shall see that questions and criticisms about Aristotle's account, in whole or in part, arise almost as soon as there are commentators on his investigations. Nevertheless, say what they will, his most prominent ancient detractors (Lucretius and Themistius spring to mind) don't succeed in trashing the whole edifice, although they do end up affecting how it will be understood and encoded by later authors. No—instead, Aristotle comes into the Arabic Middle Ages and then into Europe as an important source for both the theories of the mechanism of spontaneous generation and also, perhaps most importantly, as a source of facts about animals.[8]

A third theme, and one that I find particularly interesting in this material, is the sense of sheer wonder that our authors sometimes betray as they look at how animals come into existence and when they think hard about how that generation happens—in particular, about what its biomechanisms might be.[9] Albertus Magnus, writing in the thirteenth century, for example, conveys a palpable sense of awe to his reader as he thinks about the way the stars and the sun act together to impress forms on otherwise inert matter to create animals. This power is so strong, he tells us, that it can even impress animal forms in stone itself, where the poor creature never had a chance at viability: we now call those same stones fossils. As he reflects on the question of how precisely this works, it strikes Albertus as marvelous that the stars could have so much generative force in them as to breathe life—albeit briefly—into the very rocks of the earth. So, too, we see elsewhere human beings that are generated from mud (for biblical reasons, no less), and we can read recipes for bees and for eels and for mice. One could go on. Many of the stories are, to the modern eye, delightful, and I don't mean that in the patronizing sense of them being cute or quaint. I mean instead that we, as modern readers, by working hard to clearly understand the worlds in which our authors lived, the mechanisms by which spontaneous generation worked, and the ramifications that come from the intersections of theory and observation, we can experience something of their wonder for ourselves.

At the same time—and this is one of the experiences that, personally, keeps me in this line of work—we can see how very marvelous and wonderful it is that the world was once shaped and structured that way and that the world looked like that to some of its smartest, carefullest, and best-educated inhabitants. That they could use that knowledge not only to understand the world and their own place in it but also to manipulate that world, to make things, to carry on rich and full lives—all of this shows how very wide the possibilities are for how the world can be structured, and how richly effective many of those possibilities could be (keeping in mind, of course, the experiences and knowledge to which those structures were applied). True, those ways of understanding the world are no longer viable; they are gone now, and there are aspects of them that we may think of as well gone, but at the same time it's remarkable how accommodating the world is to a wide range of interpretations. This is not to say that we could just arbitrarily throw away modern science and replace it with medieval science and all would be well—not at all. As I have argued elsewhere, we can't really go back, in part because we have had a great number of experiences since the Middle Ages (rocketry, electricity, vaccines, microscopes) that don't fit with the old theories and that were part of the very reason we abandoned the old theories in the first place.[10]

The fourth of the five thematic strands that run through this book is the idea of complexity. In this study, we will repeatedly see how complex the theories for spontaneous generation often were; even sexual generation poses some startlingly difficult problems when it comes to explaining and understanding the nitty-gritty of the biomechanisms of reproduction. Forget everything you know about DNA, chromosomes, cell division, even of the very existence of cells in general, and now try and imagine a way to explain how a baby animal's traits—its eye and fur color, its size and even its species—can have been passed to it from the parent. What is the mechanism by which grey fur could possibly get from a parent to a pup? What component of the parent's fur worked its way down into the reproductive organs of the parent to be passed on? And in what form? Should grey fur be visible in the parent's reproductive organs or in its seed? And how did one parent's trait overrule the other parent's? Worse, think about traits that sometimes skip a generation. How is it even possible to come up with a mechanism whereby a baby can "inherit"—think about this for a moment—red hair from a grandparent, when the baby's own parents both have black hair?

As it turns out, our authors did find a way to explain these processes, both elegantly and effectively. At the same time, though, there is the further difficulty of trying to explain how these same processes—the generation of life and the determination of traits and even of species in the young—how these all work in those animals that were known not to generate sexually but that instead came into existence through the putrefaction of mud and slime.

Here the challenge is doubly interesting. Spontaneously generated animals are, certainly in many of our earliest texts, not identifiable as a class except insofar as they could simply all be said to be spontaneously generated. By the Middle Ages, a distinction between perfect animals (those that were more complex and, for the most part, generated sexually) and imperfect animals (those that were less complex and were, for the most part, generated spontaneously) would come to draw a firmer line between the classes. But in antiquity, and quite prominently in Aristotle, the class of spontaneously generated animals was simply the class of animals that had never been seen to mate and produce eggs or offspring. Their different mode of generation aside, however, spontaneously generated animals really do look and act a lot like animals that are generated sexually in many or most respects.[11] They have legs and wings, many of them, and eyes and internal organs; they seek out food, grow, and in general behave in ways that are a lot like animals that generate sexually. Indeed, sometimes very different means of generation can be found in species that are nearly identical: wasps (sexually generated), for example, and bees (generated according to a singular hermaphroditic-cum-parthenogenic system that is both asexual and caste based). All this by way of saying that the challenge in explaining spontaneous generation seems to have been to try and find explanations for the generation of life and for trait acquisition that are as widely applicable as possible—life is life, after all—and so there must be something in common between the biomechanisms by which the housefly is generated spontaneously and those by which the dragonfly is generated sexually.

Having said that, the explanations clearly can't be identical, since in one case the animals are coming-to-be by means of the actions and processes of other, already-living organisms, and in the other case the animals are coming-to-be by means of some kind of process that involves apparently inert, unliving matter. And so it is here, precisely at the conflux of similarity and difference, that we see the search for detailed and universal explanations become very interesting, very carefully thought through, and—to go back to the fifth of our thematic threads in the book—very difficult.

Difficulty, in fact, plays a few different roles in this study. In one sense, I want to keep the difficulty of a theory at least conceptually distinct from that theory's complexity. But more than that, I want to keep an eye on how our authors handle the difficulties they encounter: the difficulty of observing small animals and their organs or the difficulty of observing remote and rare animals, to pick two examples. It is also interesting, I think, when one senses our authors simply enjoying a problem *because it is difficult.* Let's face it, scientists and scholars alike, and certainly the best of both, like nothing better than a seemingly intractable problem, and the delight of working through a particularly difficult puzzle is sometimes readily apparent in an author's work—lovely moments to which the reader is sometimes privy.

Finally, difficulty shows up in another way. If we think back to Kuhn's *Structure of Scientific Revolutions* we recall that Kuhn credits revolutionary changes in scientific worldviews to the progress of a discipline through several distinct stages.[12] There is paradigm formation, when the discipline's research focus, methods, questions, and so on develop in response to a particularly influential set of earlier results, followed by the elaboration of details according to those questions and methods ("normal science"). But in the progress of normal science, anomalies inevitably accumulate, anomalies that are not answerable according to the paradigm (or worse, actually seem to contradict it). This leads to a period Kuhn calls "crisis," in which some or many individuals working in a discipline lose faith in that discipline's ability to solve certain key anomalies, and when the sheer volume of anomalies alone threatens to overwhelm those working in the labs and classrooms.[13] Now, this is not a book about scientific revolutions, and I do not happen to see things precisely in Kuhn's terms, but still, his notions of anomaly and crisis will mark for many a familiar reference point for my final emphasis on difficulty, for what are anomalies if not difficulties? Accordingly, I want, as we move through this book to keep one eye on the emergence, as well as the acknowledgment or even suppression, of this kind of difficulty.

What This Book Is and What It Is Not

This book originated as a lecture series at Johns Hopkins University, under the generous auspices of the Charles Singleton Center for the Study of Premodern Europe. In working on the lectures and their write-up, I took the mission statement of the Singleton Center to heart, which happily emphasizes interdisciplinarity and includes this statement: "The Center is united by the mission of researching the European world before the advent of modernity." As I thought about this, it struck me that the lectures presented a unique opportunity for a scholar who has done only limited work outside of his own time period to speak to, and to engage with, a wide range of specialists from different fields and from later time periods. And so an experiment gradually came to life: What would it look like to take a fully fleshed-out section of Aristotle's science, in this case his account of spontaneous generation, and then walk with it forward in time? To see how it is interpreted, responded to, challenged, reinterpreted, and ultimately how it eventually evaporates from the discussion, perhaps leaving something still tangible shaping parts of the edifice to the end.

The three lectures were originally divided between Aristotle, medieval and Renaissance responses to Aristotle, and then alternatives to Aristotle in the early modern period, ending with the debate between John Turberville Needham and Lazzaro Spallanzani, which takes place in the pages of the 1769 French translation and critique of Spallanzani's *Saggio di osservazioni microscopiche*. That skeleton is still visible in this book's version of the lectures, but certain shorter discussions and comments in the lectures needed a little more fleshing out in print than they had required or received in person. Accordingly, this book now has six chapters, some shorter, some longer, as a way of keeping the material organized in its new, slightly expanded form.

The book begins with an account of Aristotle's influential explanation of the mechanism for generation, both sexual and spontaneous, showing how inert matter can turn into animals through the action of heat. Interestingly, we shall see that Aristotle's cosmos seems to have been a place where a fairly limited number of species were generated spontaneously, certainly as compared to the worlds of many later authors, and Aristotle is very careful about trying to determine how any given species is generated. In fact, I spend chapter 2 examining Aristotle's use of observation in the biological corpus,[14] arguing not only that he is himself a careful and thoughtful observer of animals but, more to the point, that he employs deliberate rhetorical strategies to mark, for the reader, his confidence in any given observation, whether it is his or someone else's. Chapter 3 looks at ancient responses to Aristotle's account of generation, including Lucretius's atomistic and thoroughly mechanistic theory and Augustine's influential attempt to reconcile sexual and spontaneous generation with the account of creation given in Genesis. We also see in this chapter how the list of animals thought to be spontaneously generated begins to expand after Aristotle, perhaps under the influence of folk traditions. From there we look at how Aristotle's account of generation came into the Middle Ages, having been influentially and critically modified at the hands of his Arabic commentators, and how Aristotle's most prolific Latin commentator (and also one of his earliest), Albertus Magnus, incorporates and challenges his claims, both through reflection and experience. We see how Aristotle's theory is eventually united with something like Augustine's Christianized mechanism by Fortunio Liceti in the early sixteenth century (for the same theological reasons as drove Augustine). Chapter 5 offers a short interlude to reflect on a theme that has by this point been recurring in the book, which is the question of how special a thing life is—how different it is from nonliving matter. How this question is approached by our various authors has a profound impact on how they approach spontaneous generation. Finally, I turn to the beginnings of the disappearance of Aristotle from the debates on generation, as early modern thinkers start to expand the range of explanations, and as the microscope throws up new phenomena that need incorporating into those explanations.

The book ends, unusually, *in medias res*, in the middle of the debate between Needham and Spallanzani, in 1769. I do this deliberately to make a point about evidence and experiment. Most histories of biology would point out that Spallanzani responded to Needham's 1769 criticisms, and any biology textbook would see those subsequent responses as one of the key disproofs of spontaneous generation. I purposely stop short of incorporating those later arguments and experiments in order to think about how the field lay at a very specific and delicately balanced point in time before Spallanzani could respond—before anyone even knew how or if Spallanzani would or in fact could respond. And the question I want to ask is, what, in 1769, had Spallanzani and Needham really shown about boiling and sterilizing flasks?

An approach that spends more time in antiquity than later periods will mean that some structural features of this book will be a deliberate inversion of how diachronic histories are most often written, where the earliest parts of the story are the shortest, and the details and depth increase as we come closer and closer to the present. Certainly, as this project has reminded me again and again, this layout must often come about by dint of the volume of sources alone. The ancient historian in me is accustomed to having lost significantly more textual evidence than we now possess, but as we come closer to the present day, the volume of material to master becomes exponentially deeper, the lacunae finer and fewer. But by insisting on bringing forward the best version of Aristotle that I can, this book will be weighted more heavily to earlier periods and will select debates from later periods based on the earlier analysis. This is not to say that I aim to treat later authors simplistically, but that the debates and sources that I single out in the later periods are not chosen in an attempt to capture their contemporary scene in its entirety but instead to isolate threads that expand our understanding of key themes that I began developing in antiquity. This is also salutary because there are really good, detailed studies of spontaneous generation in the periods after Aristotle that already do the diachronic work in more comprehensive ways than I could ever hope to in a book of this size. For the Middle Ages there is the remarkable *Le ver, le démon et la vierge*, by Maaike van der Lugt, and for the early modern period, we are spoiled for choice: Hiro Hirai's *Le concept de semence dans les théories de la matière à la Renaissance*, for example; John Farley's *The Spontaneous Generation Controversy from Descartes to Oparin*; Elizabeth Gasking's *Investigations into Generation*; and, of course, the granddaddy of them all, Jacques Roger's 1963 *Les sciences de la vie dans la pensée française du xvii*e *siècle*.[15] For antiquity, by contrast, a synthetic history of spontaneous generation has yet to be written.

I also hope that the reader takes this book in the spirit in which it was written: as one scholar's explorations into new topics and new historical periods about which he reads but in which he has not much published. It is not, given its size, a comprehensive history of spontaneous generation as an idea. In fact, here the plural *ideas* might even be a better word choice, for the same reason that I avoid talk of "the theory of" spontaneous generation. For most of its history spontaneous generation was a fact, not a theory, about the universe: some animals just were made that way.[16] Indeed, to explain this fact, a number of different and often incompatible theories were developed that successfully fit the observational evidence, and they were embedded in epistemic contexts that determined what it meant for a theory to "successfully fit" something called "the observational evidence," likewise determined by the epistemic contexts. I will therefore talk about a theory of (spontaneous or other) generation only in the sense of it being a theory for generation, which is to say, a theory about the biomechanisms through which the phenomenon itself, whether sexual or spontaneous, is explained.

One final word about the limitations of this book. I have restricted myself to looking at theories about the spontaneous generation of animals rather than plants, and I touch on plants only when they bear directly on the discussion in a particular author. Partly this is because animals make up the lion's share of the literature on spontaneous generation, and partly for the more pragmatic reason of keeping the scope of this project under control.

A Brief Note about Latin Writing Conventions

Different standards of Latin spelling are adopted across time or by different editors, different authors, and different printers. Neo-Latin often looks different on the printed page from classical Latin, even in modern editions. Some editions of Latin texts use *u* exclusively; others switch between *u* and *v*, depending sometimes on whether it represents a semivowel or a vowel or sometimes even whether the letter occurs at the beginning or in the middle of a word. Similarly, some editions of classical texts capitalize the first word of a paragraph, some of a sentence, some of neither. In neo-Latin, it is common to see accents appear on adverbs and a few prepositions. In transcribing 1800 years' worth of shifting Latin for this book, I opted to pick one system and stick with it as consistently as possible. Readers who work on different periods may find some familiar texts or turns of phrase slightly odd to look at, but the inconsistencies were, in my opinion, just too horrible to behold. A standard system seemed much cleaner.

And so I capitalize nothing but proper nouns in Latin; I have no *quàms* about omitting accents; I use *u* and *v* following Lewis and Short; and I *assimilate* rather than *adsimilate*. I trust that the result will not be too hideous to readers of postclassical Latin. At any rate, it should cause no confusion, and that's the most important thing. Having said that, I also note by way of contrast that I could not bring myself to override the quirks of capitalization, punctuation, spelling, and italicization found in my early modern sources in English, French, and Italian, so perhaps one is doomed to some level of inconsistency after all.

Spontaneous Generation in Aristotle

ARISTOTLE, ONE OF THE world's great polymaths, is most widely known as a philosopher. His corpus includes some of our earliest and most detailed ancient texts on logic, ethics, political philosophy, metaphysics, and poetics. His physics and cosmology are a mainstay of history of science teaching—where would the "scientific revolution" be, after all, without his geocentric cosmos and four elements to rail against?[1] What has been less extensively treated, although this is beginning to be remedied, is Aristotle's biology.[2] Still, even to this day major reference works on Aristotle sometimes gloss over or skip his biology entirely, and this is unfortunate given that biology makes up the single largest topic covered in his extant corpus, occupying perhaps 25 percent of the whole.[3] To offer a sense of his significance in the history of biology, historians are fond of quoting a passage from a letter of Charles Darwin's, written to William Ogle, translator of the 1882 *Parts of Animals*:

> You must let me thank you for the pleasure which the introduction to the Aristotle Book has given me. I have rarely read anything which has interested me more; though I have not read as yet more than a quarter of the book proper. From quotations which I had seen I had a high notion of Aristotle's merits, but I had not the most remote notion what a wonderful man he was. Linnaeus and Cuvier have been my two gods, though in very different ways, but they were as mere school-boys to old Aristotle.[4]

Primarily interested in animals, Aristotle's biological corpus is foundational for ancient zoology, embryology, taxonomy, anatomy, and inheritance theory, among other subjects. In this chapter, we explore his theory of the generation of animals—how it is that different kinds of animals reproduce their species, whether their generation be sexual, parthenogenic, or (especially) spontaneous. We will see that Aristotle tries to find explanations for generation that are as widely applicable as possible, but that he is also always conscious of outlying cases, such as bees. He is very careful about the extent to which the evidence allows him to generalize, and his explanations of spontaneous generation do not map perfectly onto those for sexual generation.

Aristotle on the Birds and the Bees

At the opening of his monumental *Generation of Animals*, Aristotle outlines the different ways that animals come into being. First, he tells us, some animals are born from the mating of males and females. These he immediately limits to those animals that actually have males and females, "for this is not the case," he adds, "with all (species)," οὐ γὰρ ἐν πᾶσίν ἐστιν.[5] Since he is of necessity limiting his discussion to macroscopic organisms, this claim may at first seem surprising, but then again, there are a lot of surprising claims in store for the modern reader of this work.[6]

Of the animals that have two sexes, there is one more distinction Aristotle wishes to make, this time on the basis of whether their copulation produces animals of the same kind as themselves (lit., "homo-genus" offspring), or animals of a different kind: τὰ μὲν ἔχει τὸ θῆλυ καὶ τὸ ἄρρεν, ὥστε τὰ ὁμογενῆ γεννᾶν, τὰ δὲ γεννᾷ μέν, οὐ μέντοι τά γε ὁμογενῆ, "some [species] have males and females, such that they generate [offspring] of the same kind, others generate, but not [offspring] of the same kind."[7] This is interesting, because the animals that produce offspring different in kind (non-genus-sharing) from themselves all share another important quality that distinguishes them from other animals: they are themselves all generated spontaneously.[8]

That Aristotle should introduce us to spontaneous generation via this little detour into likeness, which is to say, into genus-sharing, and particularly his repetition of this move a little later,[9] strikes this reader as more than a little odd, but it highlights a larger point about the significance of genus and species in Aristotle's theory of reproduction and inheritance.

For Aristotle, in contrast to the modern student of taxonomy, genus and species are not fixed levels of description. For us today, to call an animal *pan paniscus* is to locate that animal in a carefully mapped set of relations to other species—ideally, to all other species—in a phylogenetic tree. For Aristotle, by contrast, genus and species are best thought of as logical relations (if we think about some of the words in English that come from the Latin genus and species, such as the adjectives *general* and *specific*, this aspect stands out quite clearly). Moreover, we can apply genus and species to anything at all, not just biology: "Nicaraguan" (to seize on the philosopher's favorite ready-to-hand example) is a species of the genus coffee, coffee is a species of the genus beverage, beverage is a species of the genus liquid, and so on. The terms are widely applicable, and they are entirely relative: the same thing can be a species at one level of analysis and a genus at another. Genus and species are also entirely relative to context: in one discussion this animal lying beside my desk may be a species of the genus dog, in another context a species of the genus quadruped, and in a third it may be a species of the genus black. When Aristotle starts talking about reproduction, genus generally refers to things that are of the same kind in some sense (bees, for example), but the senses in which he uses it can and often do shift. The root of the word *genus* in the Greek word for "family" is also often relevant, but here we must be careful; when Aristotle says humans are a species of the genus "animal," he is not pointing to a family relation in the modern genetic sense. To say with Aristotle that humans and bonobos are each a kind of animal is not to say, with Darwin, that we share common ancestors.

That Aristotle appeals to likeness of offspring and parents as his very first differentiation among animals (some animals give birth to babies of the same genus, some of a different genus) serves to highlight the centrality of genus-sharing in Aristotle's theory of generation. For the first class of animals, those that give birth to the same genus, likeness lies at the core of what Aristotle sees as one of the great *explananda* of reproduction: why do babies look like their parents and grandparents at all? How does resemblance actually get passed on from parent to child, and why does it sometimes fail? His answer to these questions takes up a considerable portion of the *Generation of Animals* and is difficult to summarize quickly, but it will be worth touching on briefly here for the light it sheds on how Aristotle understands spontaneous generation.[10]

The basic layout of Aristotle's theory of sexual generation is reasonably well known: the male's seed imposes form, which is to say, the set of qualities the baby will have, on the matter (menstrual blood) provided by the female. The female does not pass on any of what we would call genetic information at all, and the male seed does not contribute any matter to the formation of the fetus (the seed simply evaporates like fig juice does once it has curdled the milk in cheese making, says Aristotle).[11]

As for how the seed comes to carry the form that it actually carries, there were two main competing theories in antiquity. One, which we first find in the Hippocratic corpus (*On Generation/On the Nature of the Child*) says that the seed comes from all the parts of the body, and so likeness can be passed on easily since the seed is physically derived from the actual parts of the parent's body that the baby resembles. Likeness to one parent or the other comes down to a kind of competition between their two seeds—and note that on this theory, contrary to Aristotle's, both parents contribute seed, not just the father. Aristotle objects to the Hippocratic theory on a number of grounds, including the resemblance that children sometimes bear to their grandparents, whose traits cannot contribute to the seed on this model unless those traits were directly and manifestly present in the intervening parent (if likeness happens because the seed comes from the parts of the parent's body, then there is no way for a trait to skip a generation, since no part of the seed comes from the grandparent's body).[12]

What Aristotle proposes to replace this is in many ways the exact opposite of the Hippocratic theory. Rather than the seed deriving from the parts of the body, it is instead made up of the same stuff that makes up the body, which is to say, it is a by-product of the metabolism of food. Food, for Aristotle, gets concocted by the body into blood. If the organism is hot enough (which is to say, male), it will further concoct and refine some of that blood into semen. Female animals, being colder, are unable to concoct the food fully, and so for them the process never progresses beyond the production of menstrual blood. In the normal processes of nutrition in male animals, the blood that feeds the body possesses the individual animal's particular traits in potential, since that blood can and will be turned by the animal's body into its own actual nose, hair, eyes, and other features. When, therefore, that blood is further concocted and becomes semen, the semen must also possess the animal's same traits in potential, and it can thus pass those traits on to any offspring. For our purposes, the important feature here is that semen, the vector for the implantation of form in the fetus, is a residue refined from blood as part of the body's processing of food, and it contains in itself all of the male parent's traits in potential.

Now, since the only form that the male has to pass on is his own, the offspring should, if nothing else intervenes, come out both male and identical to his father (since the matter provided by the female theoretically contributes nothing to the form of the offspring).[13] But of course this is never the case, so Aristotle has to explain why a baby sometimes has her mother's nose, or his grandmother's, or grandfather's. Here he introduces two ways in which the ideal case—that of a boy exactly resembling his father—can be subverted. As the male seed acts on the female matter to enform and ensoul the new offspring, it can be "conquered" (κρατεῖσθαι) by the coldness of the mother's matter, in which case the traits of the father's form invert to their opposite, which is to say to the mother's traits (which were waiting in the matter in a way that Aristotle again calls "in potential"). In general this happens partwise: the male seed can be conquered just with respect to the sex of the baby, or just with respect to its nose and chin, for example. This offers an elegant and simple explanation for how traits pass from the two parents to the child and for how the traits of one parent can overcome those of the other. But what about a baby that has his grandfather's eyes or hair? Here the ideas of genus and species come to the fore. For Aristotle, the individual father is a specific person from a general family, which is to say that the father is a species of a genus, which is in turn a species of the genus "human." In many instances, the male seed, while not quite being conquered, is still affected in some smaller way by the matter of the menstrual blood and becomes what Aristotle calls "loosened" (λύεσθαι) by it. Where conquering had caused the trait to invert to its opposite, loosening merely causes that trait to revert from that of the species (the specific father) to that of the genus (the family more or less distantly depending on the degree of loosening, or to "human," or even, in exceptional cases, to just "animal"). This allows Aristotle to explain any degree of likeness or unlikeness between offspring and parent in any sexually reproducing animal, and as an explanation it is both economical and exhaustive—there really isn't any outcome that it is unable to explain.[14]

At the heart of Aristotle's explanatory edifice is a sharp focus on likeness to the father and its various kinds of "failures" in any individual child. Likeness is the default case, and even departures from this ideal case are parsed in terms of likeness to the mother rather than to the father, or likeness to genus rather than to species. Given his strong emphasis on likeness as both one of the most central questions of generation as well as an important part of the solution to that very question, it should occasion no surprise that when Aristotle comes to tell us that some species of animal are not generated from parents at all, but arise instead spontaneously, he comes at the point indirectly and starts by pointing out the related fact that if and when such species do reproduce sexually, they do not (and indeed, cannot) have offspring of the same kind as themselves, οὐ μέντοι τά γε ὁμογενῆ. This "comes about for good reason," he tells us, καὶ τοῦτο συμβέβηκεν εὐλόγως: if the spontaneously generated parents had offspring identical in kind to themselves, then there is no way to understand how the parents themselves could have been spontaneously generated, because if the children are really identical then they should have been identically generated. At the same time, he goes on, the offspring that these parents do have, being different in kind, must also be sterile, since if they were not they would have offspring different in kind again, who would have offspring different in kind again, ad infinitum. And infinity, says Aristotle, is ἀτελής (*atelēs*), a word combining the unsavoury if not dangerous senses of "unfinished," and "purposeless," whereas nature, on Aristotle's understanding, "always seeks an end," ἡ δὲ φύσις ἀεὶ ζητεῖ τέλος.[15]

The various senses in which *atelēs* is being used here are compounded further when we look to see what Aristotle says about spontaneous generation elsewhere. One of the more important distinctions that his opposing terms *teleios* and *atelēs* flag, a distinction with a very long historical tail in the literature on spontaneous generation, is the idea that some animals are "perfect" and some "imperfect." By the Middle Ages and the early modern period there will be considerable debate about whether spontaneous generation is limited to imperfect animals (insects, generally) or whether perfect animals (mice are a favorite example) can be generated spontaneously. Aristotle, for his part, draws the distinction slightly differently than later authors will. For him, imperfect animals seem to be defined strictly as those that do not have a likeness-relation to their parents (a point that is logically distinct from spontaneous generation). We can see this in a passage from the *Generation of Animals* where he repeats the point that some spontaneously generated animals have males and females and can reproduce sexually: ἔστι γὰρ ἔνια τοιαῦτα τῶν ἐντόμων ἃ γίγνεται μὲν αὐτόματα, ἔστι δὲ θήλεα καὶ ἄρρενα, καὶ ἐκ συνδυαζομένων γίγνεταί τι αὐτῶν, ἀτελὲς μέντοι τὸ γιγνόμενον, "there are some such animals among the insects that are generated spontaneously and have females and males, and something is born from their union, but the offspring is imperfect."[16] Notice that he does not say that the spontaneously generated parents are an imperfect kind of animal, as the later tradition will have it, simply that the offspring are imperfect—unfinished—in some sense.

What the perfect-imperfect distinction primarily points to in Aristotle is the issue of whether what the mother generates from her body (be it live offspring or egg) has at that point finished passing through all its developmental changes, becoming now like its parents. If not, then the offspring or egg, not the parent, is said to be imperfect. If, however, that offspring or egg is already "complete" in the relevant sense, then the organism is said to generate perfect eggs or perfect offspring (one wonders whether "finished" and "unfinished" might be slightly better translations, but for historical reasons I will stick with the language of perfection). Dogs, monkeys, and human beings all give birth to perfect offspring insofar as all the parts are there in (more or less) their grown-up form from the outset. When an insect comes into being as a larva, however, that larva is imperfect, in the sense that it still has radical changes to undergo before it becomes like its parents, if the larva is to do so at all. With oviparous (egg-laying) animals, the issue is slightly cloudier. Chickens "give birth" to imperfect offspring in the sense that the egg is not yet the chick, but the egg itself is said to be perfect qua egg, insofar as it does not grow or change externally once it is laid. Fish eggs, on the other hand, are imperfect even qua eggs, because they continue to grow, says Aristotle, outside of the fish.[17] As he sets up his great hierarchy of animals,[18] he says that the hottest animals give birth to live, perfect young. Slightly cooler animals lay eggs that they keep internally, but then these eggs hatch internally and they give birth to perfect, live young (dogfish, for example). One step cooler, the next animals down give birth to perfect eggs, and the next to imperfect eggs (fish, crustacea, cephalopods). Finally, the coldest animals of all give birth to larvae, which are so imperfect that they are not even eggs yet. Indeed, on Aristotle's account larvae grow outside of the mother's body until they form an egg-like shell around themselves, and then the completed animal emerges from this. That he thinks of the pupa as a kind of egg is remarkable, but he is careful to qualify the statement, never quite committing the the idea unequivocally: "the larva becomes egg-like," ᾠώδης γίγνεται ὁ σκώληξ, and "the so-called chrysalis has the function of an egg," ἡ γὰρ χρυσαλλὶς καλουμένη δύναμιν ᾠοῦ ἔχει. Elsewhere he uses similar language: οἷον ᾠὸν γίγνεται, "it becomes like an egg," or the insect emerges καθάπερ ἐξ ᾠοῦ, "just as from an egg."[19]

Since these passages are focused on animals that are generated from parents, there is no mention of spontaneous generation, and it is unclear whether the insects that generate larvae that he speaks about here include insects born spontaneously or not. It is clear that Aristotle thinks that insect species that are normally generated sexually do themselves give birth to imperfect larvae, which then metamorphose into insects like their parents (what emerges from the chrysalis is then called "perfect"), and these insects now reproduce in the same way. Similarly, he tells us that (at least some) spontaneously generated animals come into being as larvae that then pupate:

> τὰ δ᾽ ἔντομα καὶ γεννᾷ τὰ γεννῶντα σκώληκας, καὶ τὰ γιγνόμενα μὴ δι᾽ ὀχείας ἀλλ᾽ αὐτόματα ἐκ τοιαύτης γίγνεται πρῶτον συστάσεως . . . τὸν αὐτὸν δὲ τρόπον καὶ ἐπὶ τῶν ἄλλων συμβαίνει πάντων τῶν μὴ ἐξ ὀχείας γιγνομένων ἐν ἐρίοις ἤ τισιν ἄλλοις τοιούτοις, καὶ τῶν ἐν τοῖς ὕδασιν. πάντα γὰρ μετὰ τὴν τοῦ σκώληκος φύσιν ἀκινητίσαντα, καὶ τοῦ κελύφους περιξηρανθέντος, μετὰ ταῦτα τούτου ῥαγέντος ἐξέρχεται καθάπερ ἐξ ᾠοῦ ζῷον ἐπιτελεσθὲν ἐπὶ τῆς τρίτης γενέσεως.[20]
>
> Insects that generate, generate larvae, and even the ones generated not from mating but spontaneously at first come to be from a concocting of this sort . . . It is the same with all the other animals that are generated nonsexually in wool or in something else of that sort, or else in water. After they have had the form of larvae they all become unmoving, hardening a shell around themselves. Next this breaks open and the animal emerges as from an egg, perfected at its third genesis.[21]

Some of these spontaneously generated insects do seem, as adults, to reproduce—or, more properly, produce—generating larvae of their own, but different in kind from themselves, and sterile: ἐκ τούτων συνδυαζομένων γίγνεται μέν τι, οὐ ταὐτὸ δ' ἐξ οὐδενὸς ἀλλ' ἀτελές, "from their mating something is produced, but it is not the same at all, and it is imperfect."[22] Thus lice produce nits, flies larvae, and fleas a kind of grub. Whether those grubs and larvae later pupate into something else, Aristotle does not, so far as I can see, tell us. The closest we seem to come is a statement at *Generation of Animals* 715b2ff., which says:

> τούτων δ' αὐτῶν ὅσα μὲν ἐκ συνδυασμοῦ γίγνεται τῶν συγγενῶν ζῴων, καὶ αὐτὰ γεννᾷ κατὰ τὴν συγγένειαν· ὅσα δὲ μὴ ἐκ ζῴων ἀλλ' ἐκ σηπομένης τῆς ὕλης, ταῦτα δὲ γεννᾷ μὲν ἕτερον δὲ γένος, καὶ τὸ γιγνόμενον οὔτε θῆλύ ἐστιν οὔτε ἄρρεν. τοιαῦτα δ' ἐστὶν ἔνια τῶν ἐντόμων.
>
> Of the (animals), those that are generated from the union of animals of the same genus, also themselves generate according to the genus. But animals that are not generated from animals but from rotting matter, these generate, but the offspring is a different genus and is neither female nor male. Some of the insects are of this sort.

Since the offspring in this last case is sexless, it does not resemble its parents in at least that respect, but Aristotle does not here complete the thought to tell us if that offspring later pupates.[23] There are passages in which he seems to shy away from the possibility, but the degree of his commitment is far from clear, as at 723b3ff.

> ἔτι ἔνια γίγνεται τῶν ζῴων οὔτ' ἐξ ὁμογενῶν οὔτε τῷ γένει διαφόρων, οἷον αἱ μυῖαι καὶ τὰ γένη τῶν καλουμένων ψυλλῶν· ἐκ δὲ τούτων γίγνεται μὲν ζῷα, οὐκέτι δ' ὅμοια τὴν φύσιν, ἀλλὰ γένος τι σκωλήκων.
>
> Furthermore, some animals are generated neither from those of the same kind nor from those different in kind (flies and the classes of so-called fleas are of this sort). Animals are generated from these, but they are not similar in nature; they are instead a kind of larvae.

Given the degree of difference Aristotle is pointing to in these passages, where the offspring is "a different genus" and "not similar in nature," one wonders whether he supposes the larva to be its final stage of life.[24] Certainly this offspring of spontaneously generated animals never becomes like the parent and never produces its own offspring: ἐξ ὧν οὔτε τὰ γεννήσαντα γίγνεται οὔτε ἄλλο οὐδὲν ζῷον, "from these [larvae] the parent [species] is not produced nor any other animal."[25] Depending on how broadly we read γίγνεται in this passage (it means either "to be produced" or "to be born"), Aristotle is either saying these larvae do not generate or that these larvae neither generate nor pupate into something else.

The Nuts and Bolts of Spontaneous Generation

If sexually produced animals are generated through the enformation of the female matter by the agency of the male seed, how are spontaneously generated animals formed for Aristotle, in the absence of a seed to impose form on the matter of which they are made?

If we look to Aristotle's explanation of the physical makeup of the seed in sexually reproducing animals, we find that this makeup is essential to the seed's ability to enform the fetus. In an analogous way, we also see him explicitly tie the physical makeup and powers of the seed directly to the physical powers and makeup of the forces and matter that cause spontaneous generation. First, he tells us that "seed is a sharing [and note the unusual word choice in place of some version of κρᾶσις or μίξις] of pneuma and water, and pneuma is hot air," ἔστι μὲν οὖν τὸ σπέρμα κοινὸν πνεύματος καὶ ὕδατος, τὸ δὲ πνεῦμά ἐστι θερμὸς ἀήρ.[26] Shortly thereafter he elaborates:

> πάσης μὲν οὖν ψυχῆς δύναμις ἑτέρου σώματος ἔοικε κεκοινωνηκέναι καὶ θειοτέρου τῶν καλουμένων στοιχείων· . . . πάντων μὲν γὰρ ἐν τῷ σπέρματι ἐνυπάρχει ὅπερ ποιεῖ γόνιμα εἶναι τὰ σπέρματα, τὸ καλούμενον θερμόν. τοῦτο δ' οὐ πῦρ οὐδὲ τοιαύτη δύναμίς ἐστιν ἀλλὰ τὸ ἐμπεριλαμβανόμενον ἐν τῷ σπέρματι καὶ ἐν τῷ ἀφρώδει πνεῦμα καὶ ἡ ἐν τῷ πνεύματι φύσις, ἀνάλογον οὖσα τῷ τῶν ἄστρων στοιχείῳ. διὸ πῦρ μὲν οὐθὲν γεννᾷ ζῷον . . . ἡ δὲ τοῦ ἡλίου θερμότης καὶ ἡ τῶν ζῴων οὐ μόνον ἡ διὰ τοῦ σπέρματος, ἀλλὰ κἄν τι περίττωμα τύχῃ τῆς φύσεως ὂν ἕτερον, ὅμως ἔχει καὶ τοῦτο ζωτικὴν ἀρχήν.[27]

The power of every soul seems to have shared in [and he uses a word compounded on *koinos* again] a different and more divine body than

> the so-called [four] elements . . . For every [animal], what makes the seed generative inheres in the seed and is called its "heat." But this is not fire or some such power, but instead the pneuma that is enclosed in the seed and in foamy matter, and the nature in the pneuma, this being analogous to the element of the stars. This is why fire does not generate any animal . . . but the heat of the sun and the heat of animals does, not only the heat that fills the seed, but also any other residue of [the animal's] nature that may exist similarly possesses this vital principle.

There is much to unpack in this passage (as I return to it over the course of the next few pages, I will refer to it as "the vital-principle passage"). In the first instance, we should ask what the "power of every soul" is that Aristotle is talking about here. He has, for a few pages before this passage, been worrying about how and when the various parts of the soul (which he often talks about simply as various souls, in the plural) come into the developing fetus.[28] Recall that on his theory of sexual generation, the semen evaporates out, leaving no physical material behind in the developing fetus. Instead, it acts not by means of its material but by means of a generative *dynamis*, or power. Asking what the "power of every soul" is in this context is asking what causes each (part of the) soul to move from potentiality into actuality, to *become,* as the animal fetus develops and gains more faculties. Whatever this *dynamis* is, says Aristotle, it seems "to have shared in a different and more divine body than the so-called [four] elements." This is the key move that allows life to be something more than inert matter—its cause is (or at least "shares in") something beyond the four elements.

Now, in Aristotle's account of sexual generation, the adult male animal already possesses his own soul, and the seed produced by his body carries with it the same ability that the food from which it was refined has in the male parent, which is to cause the enformation and ensoulment of the new fetus in the menstrual blood:

> τοῦ δὲ σπέρματος ὄντος περιττώματος καὶ κινουμένου κίνησιν τὴν αὐτὴν καθ' ἥνπερ τὸ σῶμα αὐξάνεται μεριζομένης τῆς ἐσχάτης τροφῆς, ὅταν ἔλθῃ εἰς τὴν ὑστέραν συνίστησι καὶ κινεῖ τὸ περίττωμα τὸ τοῦ θήλεος τὴν αὐτὴν κίνησιν ἥνπερ αὐτὸ τυγχάνει κινούμενον κἀκεῖνο.[29]

> The seed is a residue and is moved with the same motion according to which the body grows when the final [refinement of] food is distributed; when [the seed] enters the uterus it organizes and moves the residue of the female with respect to that same motion by which it itself happens to be moved.

But we run into potentially serious difficulty when we try to see how this might work for spontaneously generated animals, for here there is a radical discontinuity. The material from which the animal arises is, unlike even the menstrual blood which forms the physical material of a new sexually generated animal, independent of preexisting life processes.[30]

In this case, Aristotle says, a special kind of heat will be necessary. The role of heat and pneuma here rests on his earlier discussion of the composition of semen, although in the vital-principle passage he brings in important additions to the theory. Aristotle had been worrying, a little before this passage, about the physical properties of semen, in particular, why it should be thicker when it is hot than it is when it cools. He begins that discussion by comparing the properties of semen to the properties of other known mixtures. If semen consisted of water, then it shouldn't thin as it cools, he tells us. Turning his attention to milk, which he says contains "a lot of earth," γῆς πλεῖον ἔχει,[31] Aristotle argues that, although such substances do thicken when they are heated, unlike semen they do not become "watery" when cooled. But of course water and earth aren't the only possible elements, and he next turns to mixtures of water and air. Well, not air exactly, but pneuma: "breath," that special kind of air that for many years after Aristotle would have important biological and medical properties above and beyond those of atmospheric air.[32] Pneuma, he tells us, when mixed with water produces foam, and the smaller and less visible the bubbles in a foam become, the thicker the foam gets. If we're not completely persuaded by the case of water and air (how "thick" is the foam on a wave?), Aristotle has a better example ready to hand: that of olive oil, which when beat with air—sorry, pneuma—becomes both thick and white. Clearly, he says, something analogous is happening in the composition of semen: it is an ultrafine foam composed of pneuma and something watery. This explains its color, its thickness at temperature, and more importantly, its ability to impose a living form on suitable matter.

One may be tempted to think that Aristotle is playing fast and loose by substituting pneuma for simple air in this discussion, but to be fair, I think he has a real need to try and explain processes that seem to require something more than ordinary matter—earth, air, water, fire—to explain them. Pneuma, as something transcending simple air, fills this gap nicely, even if the explanation of thickening oil may at first seem not to require elemental transcendence. Or does it? Recall his definition from 736a1 that says simply: "pneuma is hot air," τὸ δὲ πεῦμά ἐστι θερμὸς ἀήρ. It is this heat in the pneuma that is critical to its more-than-air properties, and as he tells us in the vital-principle passage, there is something special about the kind of heat that pneuma—or at least the pneuma that makes up living bodies—has. In fact, he specifically references heat in his discussion of the thickening of oil: τὸ δ᾽ αὐτὸ καὶ τὸ ἔλαιον πάσχει· παχύνεται γὰρ τῷ πνεύματι μιγνύμενον· διὸ καὶ τὸ λευκαινόμενον παχύτερον γίγνεται, τοῦ ἐνόντος ὑδατώδους ὑπὸ τοῦ θερμοῦ διακρινομένου καὶ γιγνομένου πνεύματος, "and the same thing happens to oil: it becomes thicker when mixed with pneuma. That is why the whitening becomes more compacted as the wateriness within it is separated by heat and becomes pneuma."[33] Having said that, something may have gone wrong with the text here. If pneuma is hot air, it is difficult to see on the face of it how "wateriness" could become pneuma directly. Water gets mixed with pneuma, as Aristotle said above, to form a thicker white foam, but he does not there say that the water transforms into a different element. I suggest that something has dropped out of the text. Sense can be salvaged if we substitute πνευματώδους for πνεύματος as the last word of the sentence, so that the wateriness becomes pneuma-like, which is to say, more pneuma-y.[34] Certainly that fits with the gist of the passage and has the added bonus of making better philosophical sense of it.

Either way, the point is clear: pneuma has more "heat" in it than air does, and this gives it special properties that allow the production of thick foams as the sizes of their bubbles decrease. As such foams cool, the pneuma, Aristotle says, evaporates and ὅταν ἀποπνεύσῃ τὸ θερμὸν καὶ ὁ ἀὴρ ψυχθῇ, ὑγρὸν γίγνεται, "when the heat evaporates and the air has become cold, it turns to liquid."[35] Here we see an explicit contrast between pneuma and the colder air—as the foam loses its heat we are left only with air, and the foam becomes liquid.

Now the question arises: is all hot air pneuma, or is pneuma limited to the biologically active kind of "heat"? As Aristotle tells us in the vital-principle passage, it is animal "heat" that is the generative power in the seed. The air in the seed differs from mere atmospheric air in this heat specifically, and it is akin, as he says, to ether, the element that makes up the stars, the sun, and the moon. The stars, the sun, and the moon are going to come to play a prominent role in medieval theories of spontaneous generation, but here Aristotle limits their roles, calling on their ethereal nature merely as analogies rather than as mechanisms, although he does say that there is something special about the heat of the sun that is distinct from the heat of fire and that allows it to generate animals (more on this in chapter 4).

If we now recall Aristotle's theory of the generation of the seed in the body, where the seed is a residue that is refined from blood, which is in turn refined from food, we can see him making a subtle but significant move in the last part of this passage. The heat that engenders animals "fills the seed," but it also fills "any other residue of [the animal's] nature." What might this mean? There are two possibilities, both amply attested in later traditions of spontaneous generation. First, Aristotle may be referring here to intestinal worms and other parasites that are spontaneously generated within living animals, wherein there would be some kind of residue other than seed that happens to possess a *zōtikē archē*, a vital or life-giving principle.[36] The second possibility is that he may be referring to maggots, which are engendered in the formerly living flesh of dead animals.[37] Aristotle does not dwell on either of these possibilities much in his corpus, although for what it is worth, parasites in particular become important in the later tradition, with liver flukes as one of the last bastions for spontaneous generation.

Again in the vital-principle passage, we see the appeal to "foamy matter," τὸ ἀφρῶδες. Aristotle says that the life-giving pneuma inheres in the seed "and in foamy matter." This is another clear reference to spontaneous generation, in the process of which foam and foaminess are often said to play significant roles. In addition to citing its role in the generation of testacea (an unusual genus, from the modern point of view, which includes both bivalves and snails),[38] and in the generation of the fish called the "foam-fish" (ὁ ἀφρός),[39] Aristotle makes the more general case that some kind of foam is involved at a key stage in spontaneous generation at GA 762a24. This makes sense to him for two reasons: foam is liquid mixed with pneuma or air,[40] and, drawing an analogy to sexual generation, semen is itself a foam:[41]

> αἴτιον δὲ τῆς λευκότητος τοῦ σπέρματος ὅτι ἐστὶν ἡ γονὴ ἀφρός, ὁ δ' ἀφρὸς λευκόν, καὶ μάλιστα τὸ ἐξ ὀλιγίστων συγκείμενον μορίων καὶ οὕτω μικρῶν ὥσπερ ἑκάστης ἀοράτου τῆς πομφόλυγος οὔσης.[42]
>
> The cause of the whiteness of seed is that semen is foam, and foam is white, especially that composed of the finest parts, so small that each bubble is invisible.

In fact, the likeness to sexual generation gives Aristotle significant traction, and he is able to point to the process of fetal generation from seed, as he understands it, and to show that spontaneous generation happens in much the same way. Recall that semen in males and the menstrual blood in females are both refined by the body from the digestion of food, and that these refinements are different because of the differing degrees of heat each sex possesses. But, says Aristotle, the stuff from which spontaneously generated animals are generated is also food of a sort:

> τροφὴ δ' ἐστὶ τοῖς μὲν ὕδωρ καὶ γῆ, τοῖς δὲ τὰ ἐκ τούτων, ὥσθ' ὅπερ ἡ ἐν τοῖς ζῴοις θερμότης ἐκ τῆς τροφῆς ἀπεργάζεται, τοῦθ' ἡ τῆς ὥρας ἐν τῷ περιέχοντι θερμότης ἐκ θαλάττης καὶ γῆς συγκρίνει πέττουσα καὶ συνίστησιν. τὸ δ' ἐναπολαμβανόμενον ἢ ἀποκρινόμενον ἐν τῷ πνεύματι τῆς ψυχικῆς ἀρχῆς κύημα ποιεῖ καὶ κίνησιν ἐντίθησιν.[43]
>
> For some, the earth and water are food, for others, it is the things that [grow] from earth and water, so that in this way the heat that animals have gets worked up from their food, and in this case [that of spontaneously generated animals], the heat of the season in the environment is combined, ripening, and it sets them. And this ingathering or separating of the soul-principle in the pneuma makes an embryo and starts its motion.

By pointing out that it is earth and water from which animals are generated spontaneously, and that the earth and water are themselves food for these (and other?) animals, Aristotle is further extending his analogy to sexual generation, which is, ultimately, a complex by-product of the processes of nutrition. Of course, the process is here said to require a second actor, the heat of the air, flagged prominently in a number of passages, in order to complete the process of what sounds like an initial "skinning over" of the animal, with verbs referring to "setting," "gathering in," and "separating" the embryo. This would seem to be analogous to a process Aristotle describes in the formation of a sexually generated fetus, where the initial moment at which the thing changes from a mixture of semen and blood into a living fetation and begins to take nutrition into itself (and at the same time to actualize the nutritive soul that had been in the semen in potential) is called its χωριζόμενον, its "separation."[44] This must be, for Aristotle, the moment that the animal comes into existence as a new living thing, given the definitional equivalency between living things and possessors of actualized souls.[45]

Turning back to the role of the heat from the sun in the vital-principle passage, there are two curious passages that seem to gesture at it indirectly, one from the *History of Animals* and one from the *Generation of Animals*, where Aristotle curiously emphasizes the role of water coming from the sky for spontaneous generation:

> ὅσα δὲ μήτε παραβλαστάνει μήτε κηριάζει, τούτων δὲ πάντων ἡ γένεσις αὐτόματός ἐστιν. πάντα δὲ τὰ συνιστάμενα τὸν τρόπον τοῦτον καὶ ἐν γῇ καὶ ἐν ὕδατι φαίνεται γιγνόμενα μετὰ σήψεως καὶ μιγνυμένου τοῦ ὀμβρίου ὕδατος.[46]

> With regard to such [testacea] as do not propagate by sending up shoots or by "honeycombing,"[47] their genesis is spontaneous in all cases. All animals that come to be in this way, both on land and in the water, appear to be generated by putrefaction that has also been mixed with rain water.

Here Aristotle emphasizes that he is not talking about just any water, but about rain, which is to say, water that has fallen from the sky. He even more sharply reinforces this emphasis by calling on the sky directly in the *History of Animals*, in a discussion of the so-called foam-fish, ὁ ἀφρός:

> φαίνεται δ' ἐν μὲν τόποις τοιούτοις καὶ εὐημερίαις τοιαύταις, γίγνεται δ' ἐνιαχοῦ καὶ ὁπόταν ὕδωρ πολὺ ἐξ οὐρανοῦ γένηται, ἐν τῷ ἀφρῷ τῷ γιγνομένῳ ὑπὸ τοῦ ὀμβρίου ὕδατος, διὸ καὶ καλεῖται ἀφρός.[48]

> [This fish] appears in places like those described and in the aforementioned good weather, but it also occurs in some places also whenever there has been much water from the sky, [coming-to-be] in the foam that is formed by the rain water, which is why it is called the foam-fish.

Although he does not say so directly, it may well be that the proximity of the rain water to the sun had infused it with a special kind or special degree of heat for generation that distinguishes rain from the ordinary water in which the animals live. This is complicated, however, by Aristotle's earlier assertion that the foam-fish, in regions around Athens at least, is normally generated "when the ground is heated up after the weather has become nice," ὅταν εὐημερίας γενομένης ἀναθερμαίνηται ἡ γῆ,[49] with no mention of rain at all, but a hint again at the implied role of the sun in heating the ground.

If we were to complete the analogy with sexual generation and say that the sun acts as the father, implanting the form on the material of the earth (mother), we would be going a little further than Aristotle himself actually goes. There are hints all over the place that he has something like this in mind as a loose comparison, not least because of a passage at the very beginning of the *Generation of Animals,* where he gives us his very definitions of male and female:

> ἄρρεν μὲν γὰρ λέγομεν ζῷον τὸ εἰς ἄλλο γεννῶν, θῆλυ δὲ τὸ εἰς αὐτό· διὸ καὶ ἐν τῷ ὅλῳ τὴν τῆς γῆς φύσιν ὡς θῆλυ καὶ μητέρα νομίζουσιν, οὐρανὸν δὲ καὶ ἥλιον ἤ τι τῶν ἄλλων τῶν τοιούτων ὡς γεννῶντας καὶ πατέρας προσαγορεύουσιν.[50]

> We call a male animal that which generates in another, and the female that which generates in itself. For this reason, with regard to the cosmos, the nature of the earth is as a female and people call her mother, whereas the sky and sun and anything else of that sort is the generator and people address him as father.

Now it may be tempting to interpret this as an ascription of the roles of the sun and the earth in spontaneous generation, but we should resist the urge to do so for two reasons. One is that Aristotle seems just to be throwing this analogy out, off the cuff, and he does not link it specifically to spontaneous generation; instead, he seems to be speaking more generally about the metaphorical language people use every day for the growth of all kinds of plants and animals. Secondly, as we look back at the complex mechanisms we have seen for spontaneous generation, there is never quite a one-to-one mapping with the mechanisms of sexual generation. Instead, we see comparisons and analogies drawn, but always with small caveats, always incompletely. For example, in modern scholarship on the significance of spontaneity for Aristotle's more general treatment of nature as teleologically driven, it is frequently pointed out that the only way to understand the regularity of spontaneous generation—the fact that maggots are formed in meat but mussels in ocean sand—is to assume that the form of the animal is somehow impressed by the matter, which would be to say the mother, a very different model than Aristotle has for sexual generation.[51]

In fact, it may be that the matter is somehow both the father and mother, activated, but not enformed, by celestial heat or the heat of putrefaction, as we see in perhaps the most famous passage on spontaneous generation from the *Generation of Animals*:

> γίγνονται δ᾽ ἐν γῇ καὶ ἐν ὑγρῷ τὰ ζῷα καὶ τὰ φυτὰ διὰ τὸ ἐν γῇ μὲν ὕδωρ ὑπάρχειν ἐν δ᾽ ὕδατι πνεῦμα, ἐν δὲ τούτῳ παντὶ θερμότητα ψυχικήν, ὥστε τρόπον τινὰ πάντα ψυχῆς εἶναι πλήρη· διὸ συνίσταται ταχέως ὁπόταν ἐμπεριληφθῇ . . . αἱ μὲν οὖν διαφοραὶ τοῦ τιμιώτερον εἶναι τὸ γένος καὶ ἀτιμότερον τὸ συνιστάμενον ἐν τῇ περιλήψει τῆς ἀρχῆς τῆς ψυχικῆς ἐστιν. τούτου δὲ καὶ οἱ τόποι αἴτιοι καὶ τὸ σῶμα τὸ περιλαμβανόμενον.[52]

> Animals and plants are generated in the earth and in water because there is water in earth, and in water there is pneuma, and in all pneuma there is soul-heat, such that, in a certain sense, everything is full of soul. This is why it comes together quickly when it has been enclosed . . . The differences between there being generated a more honorable or a less honorable species are in the encasement of the life principle. And the causes of this are the locations and the matter that is enclosed.

Species determination, then, seems to be entirely due to the composition of the material in which spontaneous generation occurs. This is why, Aristotle tells us, particularly earthy soil generates testacea, for example.

Thus for Aristotle, spontaneous generation is a process of the activation in matter of the potential for life that inheres in all pneuma. This pneuma is akin to the divine substance of the stars. When it is intermixed in some way with matter sufficiently and in tiny bubbles throughout, and when it is acted on by the heat imparted to the atmosphere or the water by the sun, we will witness the spontaneous generation of one species or another, depending on the material. As we have seen, this theory is to some extent modeled on Aristotle's theory of sexual generation, but in some ways it is necessarily independent. It is notable that Aristotle carefully limits his list of species that are spontaneously generated, and, as we will continue to see in chapter 2, he is exceedingly careful about canvassing and clarifying his varying degrees of confidence in his different pieces of evidence for the reader.

CHAPTER TWO

Aristotle and Observational Confidence

ARISTOTLE IS WIDELY recognized as an astute and careful observer of the natural world.[1] However, for all that Aristotle's own observational care and innovation are frequently commented upon, we have no systematic and exhaustive picture of how Aristotle deploys and epistemically qualifies observation statements—his own or other people's—in the biological corpus.[2] What we do see is that Aristotle adopts a number of rhetorical strategies for presenting observation claims, and that the ways in which he does so carefully (and quite consistently) qualify for his reader the degrees of confidence that he himself has in the various claims. This question of Aristotle's confidence is different from asking about the quality or accuracy of Aristotle's observations;[3] instead, it draws attention to the pains Aristotle takes to reflect on or to eliminate possibilities when things are ambiguous, and specifically to signal his own doubts, hesitations, and qualifications—or alternatively, to signal his confidence in—individual observation claims.

Aristotle's biology, it bears saying, relies rather heavily on observation claims about how one organism or another functions, how it moves, how it is constructed, how it compares to other organisms. Of course, physics, alchemy, astronomy, and any number of other scientific subjects also rely on observation in antiquity, but what makes ancient biology different from these other subjects is the degree to which its observations come from third-party sources, as well as the degree to which it is what we might call a science of special cases. What I mean by this is that there are considerably more species of animals or plants than there are kinds of stars in antiquity or kinds of physical matter.[4] Heavy objects behave the same way in

Athens as they do in Alexandria; a cup of drinking water has, for all intents and purposes, the same properties on land as it does on board a ship. Earth falls, fire goes up, magnets attract, always and everywhere. But in biology, the deer outside of Athens may be significantly different from the deer outside of Alexandria, which may be significantly different again from whatever corresponds to a deer in the mountains of India. There is a wide range of animals here and in other parts of the world that needs accounting for and explaining, no matter how strange or unique the animals seem to be—elephants, hyenas, and who knows what else—and for many of these Aristotle is going to have to rely on observations taken by people other than himself. True, ancient astronomers and other writers on the natural world rely heavily on observation reports from others, but an eclipse observed by someone in Babylon 500 years ago is still fundamentally an eclipse; the only issues are how well it is timed, dated, and described. A goat observed at Babylon, on the other hand, may have features and properties entirely unheard of in Macedonian goats. Animals at sea are entirely different from those on land (except when they are not, of course). One could go on. In short, the raw data of Aristotle's biology is significantly more complex, much messier, and frequently more local than the raw data in other parts of his oeuvre. We should look, then, at precisely how Aristotle handles that data, those observation claims.

Combing through Aristotle's biological corpus for observation reports, for words and phrases denoting things that "have been seen" (at some point) or "were observed" (on particular occasions), as well as phrases denoting testimony ("they say," "it is said that," "so-and-so says"), we find that observation claims in Aristotle fall under three broad epistemological headings: things he believes to have been accurately observed (sometimes clearly by himself, sometimes clearly by others, sometimes not clearly ascribed), things he believes to have been poorly observed or gullibly reported (always by others), and observations (by himself or others) for which he wishes to qualify his certainty in one way or another. In this last case, he qualifies his claims using three primary rhetorical strategies: (a) telling us that an observation has been reported by someone else (leaving the reader to judge its plausibility), (b) telling us that an observation is or was "difficult" to make or otherwise uncertain, or else (c) telling us that an observation seems secure, insofar as no counterinstances have yet been observed, thereby implicitly carving out an obvious space for the possibility that things may one day change, and with them our theories about the animals or processes involved.

Simple Facticity

Aristotle's general practice in reporting an observation is to simply put that observation in the passive voice: "it has been seen that . . . ," most commonly in the Greek perfect tense (e.g., ὤπται) or else in the aorist tense (e.g., ὤφθη). Modern scholars have often taken many of these reports to be Aristotle's own observations,[5] and it is certainly possible, and even quite tempting, to think that they are his own, although there are also instances where it is clear that Aristotle is likely not himself the observer, either because the observations in question are well attested in earlier authors (Herodotus is a favorite) or because it is implausible that Aristotle himself, or in some cases any single individual working alone, could have made the observations in question. But if we cannot always be certain that Aristotle himself made some specific observation that he reports, one thing we can tease out from many of his observation statements is that, for his own part, Aristotle has confidence in their veracity, and this is significant for how he uses the observations in his work.

In some instances, although perhaps fewer than modern readers might wish for, Aristotle explicitly tells us that he has personally made a given observation. So, for example, he tells us that the hearts of animals do not generally contain bone: ἔστι δ' ἀνόστεος πάντων ὅσα καὶ ἡμεῖς τεθεάμεθα, πλὴν τῶν ἵππων καὶ γένους τινὸς βοῶν, "[the heart] is boneless in all animals that we ourselves have observed, except in horses and one kind of cattle."[6] Here the claim to autopsy is both clear and emphatic. Greek does not normally need to deploy personal pronouns as grammatical subjects (*I* and *we*, for example) since the doer of an action is already signaled in the verbal ending (-άμεθα, in this case). When authors do so, it is because they want to emphasize their own or another actor's involvement a little more strongly than the default wording would offer (hence my use of the reflexive "we ourselves," strengthened in the Greek with an adverbial καί). Aristotle leaves no room for doubt here. But at the same time, he is subtly qualifying his knowledge claim, as we will see him often do in later examples, by pointing out that the experiential basis to which the claim is being pegged is inherently limited: the phrase "in all the animals that we ourselves have observed" clearly leaves open for the reader the possibility that there are animals still out there, perhaps not native to Greece, or at any rate ones not yet brought to Aristotle for dissection, that also have bones in their hearts. He is saying: there are at least these two exceptions to the general rule about hearts, but there may be more.

More confident is a passage in the *Generation of Animals* where Aristotle rules out the theory that, in cases of multiple births, males are formed in one part of the womb and females in another:

> ἔτι δὲ γίγνεται δίδυμα θῆλυ καὶ ἄρρεν ἅμα ἐν τῷ αὐτῷ μορίῳ πολλάκις τῆς ὑστέρας, καὶ τοῦθ' ἱκανῶς τεθεωρήκαμεν ἐκ τῶν ἀνατομῶν ἐν πᾶσι τοῖς ζῳοτοκοῦσι, καὶ ἐν τοῖς πεζοῖς καὶ ἐν τοῖς ἰχθύσιν.[7]
>
> Moreover, both male and female twins are often formed at the same time in the same part of the womb, and we have observed this sufficiently in dissections of all the vivipara, in both land animals and fish.

Here Aristotle shows no doubt whatsoever about his conclusion, drawn from what he implies is a rich store of observation "in all the vivipara," and the evidence that he has collected is, he tells us explicitly, "sufficient" in quantity and quality to make the case.[8] In contrast to the previous observation claim, with its mildly emphatic "we ourselves have observed" (καὶ ἡμεῖς τεθεάμεθα), in this instance Aristotle uses the simpler and more common wording, which leaves the personal pronoun unstated (τεθεωρήκαμεν), while still laying explicit claim to personal observation.

Elsewhere we see a claim to autopsy, though with perhaps less confidence in his conclusions, when he asserts that, of all the organs in the fetus, the heart is formed first. He says: τοῦτο δὲ δῆλον ἐξ ὧν ἐν τοῖς ἐνδεχομένοις ἔτι γιγνομένοις ἰδεῖν τεθεωρήκαμεν, "this is clear from all [the cases] we have seen where it is possible to see them coming-to-be."[9] Again we see his use of the first-person-plural verb, but it is conjoined with a frank admission that in some, or even many cases, there are difficulties of observation that may make certainty elusive.

Beyond these, there are not a great number of instances in the biological corpus where Aristotle is quite so explicit about observations that he himself has made, although there are a host of observations that most commentators are happy to ascribe to him without his explicit confirmation (and indeed, I largely follow them). In some of these cases, the evidence is reasonably strong and direct that he himself was the observer, but in others the conclusion is admittedly less definite. One such instance we find in a long passage beginning at HA 511b11, where he criticizes in great detail the writings of his predecessors Syennesis of Cyprus, Diogenes of Apollonia, and Polybus (said in some sources to be the son-in-law of Hippocrates) on the anatomy of the blood vessels. He opens his criticism as follows:

> αἴτιον δὲ τῆς ἀγνοίας τὸ δυσθεώρητον αὐτῶν. ἐν μὲν γὰρ τοῖς τεθνεῶσι τῶν ζῴων ἄδηλος ἡ φύσις τῶν κυριωτάτων φλεβῶν διὰ τὸ συμπίπτειν εὐθὺς ἐξίοντος τοῦ αἵματος μάλιστα ταύτας . . . ἐν δὲ τοῖς ζῶσιν ἀδύνατόν ἐστι θεάσασθαι πῶς ἔχουσιν, ἐντὸς γὰρ ἡ φύσις αὐτῶν.[10]

> The cause of their ignorance is the difficulty of observing [the blood vessels]. In animals that are dead, the nature of the most important vessels is unclear because these collapse completely as soon as the blood leaves them . . . and in the living it is impossible to observe how they are situated because their position is internal.

He then quotes long passages from each author,[11] summarizes the ideas of a few unnamed others, and dismisses them all on the grounds that they are collectively mistaken in their view that the source of the blood vessels is ultimately in the head.

> χαλεπῆς δ᾿ οὔσης, ὥσπερ εἴρηται πρότερον, τῆς θεωρίας ἐν μόνοις τοῖς ἀποπεπνιγμένοις τῶν ζῴων προλεπτυνθεῖσιν ἔστιν ἱκανῶς καταμαθεῖν, εἴ τινι περὶ τῶν τοιούτων ἐπιμελές.
>
> ἔχει δὲ τοῦτον τὸν τρόπον ἡ τῶν φλεβῶν φύσις . . .[12]

> Observation being, as I said before, difficult, it is only in those animals that have been strangled that it is possible to learn sufficiently, if one is concerned about such things.
>
> And the nature of the blood vessels is as follows . . .

At which point Aristotle gives a long, careful, and detailed description of the layout of the blood vessels in animals. He discusses their positions and sizes as though for a generic animal, occasionally with more specific remarks on the differences between some particular feature or other in small animals versus large, or between males and females, or in quadrupeds versus birds versus fish. Since Aristotle has earlier dismissed his predecessors' accounts and cautioned his reader about the proper way to kill animals for dissection, it is a natural conclusion that his description of the blood vessels is rooted in his own very careful observation, even if he does not say it in so many words. Since he discusses distinctions between different kinds of animals and offers advice about which are best for observation, one also gets the impression that his experience in this kind of dissection is extensive, an impression further confirmed by his tantalizingly common references to a sadly long-lost work of his called *The Anatomies*.[13]

Other passages occasionally also establish facticity by hinting strongly at the author's personal experience. Take, for example, Aristotle's discussion of the general anatomy of the animal heart: πᾶσι δ' ὁμοίως καὶ τοῖς ἔχουσι καὶ τοῖς μὴ ἔχουσι τοῦτο τὸ μόριον [sc. τὸ στῆθος] εἰς τὸ πρόσθεν ἔχει ἡ καρδία τὸ ὀξύ· λάθοι δ' ἂν πολλάκις διὰ τὸ μεταπίπτειν διαιρουμένων, "the point of the heart is toward the front in all animals, both those that have [a chest] and those that do not, but this can often escape one's notice because of a change during dissection."[14] Here the cautionary note for the reader about the care needed to make the observation correctly would seem to be sounding a clear note of personal experience on Aristotle's part. Even if he is not quite explicit about personal autopsy, we at least see him setting out the observation without doubt or hesitation.

The series of detailed observations with which he describes the development of the chick in its egg at HA 561a4-562a22 are also widely taken to be his own, although as usual, all that he offers is a clinical description of the parts and their order of emergence, with an occasional interjection that this or that feature becomes "clear" (δῆλος, φανερός) or that some feature "appears" (ἐπισημαίνει) at a certain time.[15]

One observation that Aristotle reports with full certainty and that sounds like it may well be his own has caused some consternation among later commentators, as it is entirely unclear what Aristotle thinks he or his source was seeing: αἱ δ' ἔγχελυς, he tells us, γίγνονται ἐκ τῶν καλουμένων γῆς ἐντέρων, ἃ αὐτόματα συνίσταται ἐν τῷ πηλῷ καὶ ἐν τῇ γῇ τῇ ἐνίκμῳ. καὶ ἤδη εἰσὶν ὠμμέναι αἱ μὲν ἐκλυόμεναι ἐκ τούτων, αἱ δ' ἐν διακνιζομένοις καὶ διαιρουμένοις γίγνονται φανεραί, "eels . . . are born from the so-called earth-worms [lit., earth's guts] which spring up spontaneously in clay and in damp earth. [Eels] have now been seen emerging from these, and they become visible when [the worms] are cut open and dissected."[16] This observation, reported with perfect confidence by Aristotle, is a key piece of the evidence for him that eels are spontaneously rather than sexually generated. Nevertheless, given that no part of the difficult-to-observe reproductive cycle of the eel involves anything worm-like happening on shore, it is hard to know what the actual observation might have been in this instance. That being said, we note that Aristotle presents the observation as though it were entirely unproblematic.

It is based on confident passages such as these that scholars have come to believe that Aristotle made many, if not most, of the dissections and many of the other observations described in his biological corpus himself. I agree that the conclusion is often more than tempting, but of course, he generally does not claim personal autopsy explicitly, and caution may be called for on a case-by-case basis. What is, however, abundantly clear from such passages—and this is for our purposes the more important point—is that Aristotle is signaling his complete confidence in these observations. For him, these are unproblematic facts, put to the reader simply, straightforwardly, and without reservation.

In other instances Aristotle shows every sign of confidence in an observation he reports, but for various reasons the reader may well suspect that he is repeating testimony given to him rather than reporting on his own experience. When he discusses, for example, multiple births in humans, he says: πλεῖστα δὲ τίκτεται πέντε τὸν ἀριθμόν· ἤδη γὰρ ὦπται τοῦτο καὶ ἐπὶ πλειόνων συμβεβηκός. μία δέ τις ἐν τέτταρσι τόκοις ἔτεκεν εἴκοσιν· ἀνὰ πέντε γὰρ ἔτεκε, "the greatest in number born is five. And this event has by now been seen on rather a number of occasions. One woman gave birth to twenty children in four births, for she birthed in fives."[17] Now, it is possible that this woman's four multiple births constitute the entirety of the "number of occasions" (ἐπὶ πλειόνων) that Aristotle says back him up, but to my ear the Greek sounds a little more plentiful than this (that καὶ helps). I suspect Aristotle had other reports in mind, meaning that, given the rarity of quintuplets, his ὦπται, "it has been seen" likely includes testimonial evidence (and indeed, there is no guarantee that Aristotle himself had actually even seen the one very fecund woman he describes).

Similar hints that testimony may lie behind a given observation claim litter the corpus. For example: πολλαχοῦ τεθεωρῆσθαι, "it has been observed," he says, "in many places" that hedgehogs move the entrances to their burrows in response to changes in wind direction.[18] It certainly sounds as if Aristotle is alluding to other observers here besides himself. Even clearer is an observation he records about hyenas. The hyena, he says, "is seen," ὤπται, to have only one set of genitals, ἐν ἐνίοις γὰρ τόποις οὐ σπάνις τῆς θεωρίας, "in some places there is no shortage of this observation."[19] Here again, Aristotle sounds as if he is alluding to observations made by others, and his "in some places" is more plausibly taken as referring to foreign locales than to private collections he may have personally visited. Nevertheless, it is clear from the references to ubiquity that he wants his reader to read these reports from elsewhere as straightforwardly factual.

Testimony (qualified)

Aristotle is often very clear about the source of his testimonial observations, and he uses phrases that mark out the observer explicitly as someone other than himself, thereby implicitly qualifying his certainty. In many instances he is more or less vague about who, precisely, these other observers are, using a bald "they say," φασίν, while at other times he is a little more specific: "fishermen say," or "shepherds say."

Instances of the simple use of "they say" in the biological corpus include stories that Aristotle clearly finds unproblematic, such as when he tells us of dolphins: διαπορεῖται δὲ περὶ αὐτῶν διὰ τί ἐξοκέλλουσιν εἰς τὴν γῆν· ποιεῖν γάρ φασι τοῦτ' αὐτοὺς ἐνίοτε ὅταν τύχωσι δι' οὐδεμίαν αἰτίαν, "why they beach themselves on land is a puzzle, for they say that [dolphins] sometimes do this whenever they may happen to for no reason."[20] What we can see is that he thinks the fact itself to be fairly straightforward, setting it out as a known puzzle about the species, while at the same time distancing himself from any potentially literal reading of "for no reason," insofar as he implies that there must be some reason for the behavior if the question is in fact to be a puzzle.

In other instances it is less clear whether Aristotle believes in what "they say" or not, as when he tells us that people say that salamanders put out fires by crawling through them, or when he reports that "they say" that a certain Syracusan drunkard used to bury fresh eggs under his blanket and not stop drinking until they hatched.[21] He also attributes to some unstated "them" that fact that the *phykis* is the only fish to build nests and that elephant handlers give olive oil to elephants to expel any iron splinters they may have picked up.[22] Elsewhere, Aristotle prefaces a story about elephants with "they say that in India . . . ,"[23] reminding us that, as with hyenas a moment ago, some facts or curiosities that Aristotle wants to record concern animals to which he simply has no access because of geographical distance. Similarly, we sometimes see him report facts from observers "in Libya," or "in Crete," or "around Lake Maeotis," to take a few examples.[24] Finally, Aristotle sometimes flags his certainty or uncertainty in testimonial observations more explicitly, by using phrases like "it has been observed often," or at the other end of the spectrum, "some people say."

We can also see him sometimes accepting part of a story that "they say" but rejecting another part (possibly reported by a different "they"). Thus, when Aristotle tells us that "some [fish] are generated out of mud and out of sand," ἔνιοι καὶ ἐκ τῆς ἰλύος καὶ ἐκ τῆς ἄμμου γίγνονται, which he reports as an unproblematic fact, the evidence he offers is testimonial:

> ἐν τέλμασιν ἄλλοις τε, καὶ οἷον περὶ Κνίδον φασὶν εἶναί ποτε, ὃ ἐξηραίνετο μὲν ὑπὸ κύνα καὶ ἡ ἰλὺς ἅπασα ἐξηρεῖτο, ὕδωρ δ᾽ ἤρχετο ἐγγίγνεσθαι ἅμα τοῖς πρώτοις γιγνομένοις ὄμβροις. ἐν τούτῳ δὴ ἰχθύδια ἐνεγίγνετο ἀρχομένου τοῦ ὕδατος. ἦν δὲ κεστρέων τι γένος τοῦτο, ὃ οὐ γίγνεται ἐξ ὀχείας, μέγεθος ἡλίκα μαινίδια μικρά· ᾠὸν δὲ τούτων εἶχεν οὐδὲν οὐδὲ θορόν. γίγνεται δὲ καὶ ἐν ποταμοῖς ἐν τῇ Ἀσίᾳ ὅπου διαρρέουσιν εἰς τὴν θάλατταν, ἰχθύδια μικρά, ἡλίκοι ἑψητοί, ἕτερα τὸν αὐτὸν τρόπον τούτοις.[25]
>
> [This happens] in different ponds, and they say one near Cnidus is sometimes like this. It dried up at the time of the Dog Star and all the mud hardened. But then water came when the first rains arrived. Little fish appeared in it as soon as the water started, and this was a species of mullet which is not generated by mating. Its size was like a small sprat, and in them there was neither egg nor milt. Different small fish are generated in the same way as these in the rivers of Asia where they flow into the sea.

Notice the little details that Aristotle throws in for verisimilitude: the time of year, the location, the size of the fish, and even the matter-of-factness that the species is (already?) known not to be generated by mating. But above and beyond these, he includes one of his favorite and most convincing observations for proving that eels are not sexually generated, repurposed here for our spontaneously generated mullets: none of them had eggs or milt. One gets the impression from how it is reported that Aristotle is almost looking over the shoulder of his source. Who this may have been is left unsaid, although it must have been someone interested in exactly the question of how these fish were generated. Who else, after all, would care enough to notice all the details? As he continues this discussion, however, Aristotle's tone shifts, for not everything that people say is equally reliable: ἔνιοι δὲ καὶ οὕτως φασὶ τοὺς κεστρεῖς φύεσθαι πάντας, οὐκ ὀρθῶς λέγοντες· ἔχουσαι γὰρ φαίνονται καὶ ᾠὰ αἱ θήλειαι αὐτῶν καὶ θορὸν οἱ ἄρρενες. ἀλλὰ γένος τί ἐστιν αὐτῶν τοιοῦτον, ὃ φύεται ἐκ τῆς ἰλύος καὶ τῆς ἄμμου, "some say that all mullets are generated in this way, but they say so incorrectly, for the females are observed with eggs and the males with milt. But there does exist a species of them like that, which is generated from mud and sand."[26]

In other instances, Aristotle tells us who his sources are, but this information comes in tandem with what we might call a conditional appeal to authority. This occurs where Aristotle reports the testimony of professionals of one sort or another. So fishermen say that certain parasitic crabs come into existence with their hosts; women who hunt for drugs say that a certain medicinal substance is difficult to obtain from mares; shepherds say that wind direction has an effect on the sex of lambs, and beekeepers say—well, beekeepers say rather a lot of things.[27] But we need to keep in mind that these appeals to professional lore are inherently conditional. On the one hand, the reader recognizes that professionals have an intimate knowledge of the tools and objects of their day-to-day labor and so will often be able to tell us something that we would not otherwise know about the animals with which they work. At the same time, the ancient reader would be acutely aware that these fishermen, shepherds, and women who collect medicines were almost certainly uneducated and illiterate, and so perhaps less than completely reliable on a number of fronts (do they understand the questions or their observations properly? Are they capable of reporting carefully or of noticing relevant potential counterevidence? Can they be trusted to tell the truth?). Indeed, Aristotle explicitly acknowledges the problem in a discussion of the mistaken belief that female fish swallow the milt of the male. He tells us that those who hold this view have failed to carefully observe certain details (οὐ κατανενοηκότες ἔνια λέγουσιν οὕτως, "some say this without having thought it through"),[28] and then a little later explains that πολλοὺς λανθάνειν καὶ τῶν ἁλιέων, οὐθεὶς γὰρ αὐτῶν οὐθὲν τηρεῖ τοιοῦτον τοῦ γνῶναι χάριν, "[the facts] escape the notice even of many fishermen, for none of them observes [mating] for the sake of study."[29] They may see it incidentally, but their observations are colored by their having no reason to look closely at the relevant details or even perhaps to know what the relevant details might be. The problem is exacerbated, he tells us elsewhere, by the speed with which fish mate: ἡ δ' ἀληθινὴ σύνοδος τῶν ᾠοτόκων ἰχθύων ὀλιγάκις ὁρᾶται διὰ τὸ ταχέως ἀπολύεσθαι, "the real mating of egg-bearing fish is seldom seen due to their speedy completion."[30] "Still," he immediately reminds us, care will be rewarded, for "even in this case copulation has been observed," ἐπεὶ ὦπται καὶ ἡ ἐπὶ τούτων ὀχεία.

Whether or not Aristotle believes in any individual claim with more or less certainty, we can see him clearly distancing this testimony from his own observation and experience by the use of formulae to attribute these observations unambiguously to others. The reader can see exactly where the evidence stands for these claims.

Degrees of Uncertainty

Having made the admission that sometimes tradespeople will miss important facts, Aristotle still frequently (and necessarily) relies on their expertise, although he often handles it with care. One particularly cautious deployment of professional knowledge occurs in contexts in which Aristotle is himself unsure what to make of the evidence, as he tells us explicitly. Here the uncertainty finds one of two sources: either the evidence is rare and therefore difficult to interpret, or else it is ambiguous. Two such instances occur in close proximity to each other and involve professional fishermen or sailors.[31] In the first example, Aristotle's uncertainty stems from the rarity of the phenomena:

> ἔστι δ' ἔνια ζῷα περιττὰ καὶ ἐν τῇ θαλάττῃ, ἃ διὰ τὸ σπάνια εἶναι οὐκ ἔστι θεῖναι εἰς γένος. ἤδη γάρ φασί τινες τῶν ἐμπειρικῶν ἁλιέων ἑωρακέναι ἐν τῇ θαλάττῃ ὅμοια δοκίοις, μέλανα, στρογγύλα τε καὶ ἰσοπαχῆ· ἕτερα δὲ καὶ ἀσπίσιν ὅμοια, τὸ μὲν χρῶμα ἐρυθρά, πτερύγια δ' ἔχοντα πυκνά· καὶ ἄλλα ὅμοια αἰδοίῳ ἀνδρὸς τό τ' εἶδος καὶ τὸ μέγεθος, πλὴν ἀντὶ τῶν ὄρχεων πτερύγια ἔχειν δύο, καὶ λαβέσθαι ποτὲ τοιοῦτον τοῦ πολυαγκίστρου τῷ ἄκρῳ.[32]
>
> There are some strange animals in the sea which, because of their rarity, cannot be assigned to a genus. For instance, some experienced fishermen say that they have seen in the sea animals like planks, black, round, and of uniform thickness; others like shields, red in color, having many fins; and still others like a man's genitals in both size and shape, except that in place of testicles they have two fins, and [they say that] one was once caught on the hook of a longline.

Here Aristotle makes it clear that he is not entirely sure what to make of these animals, but not that he is unsure what to make of the truth-value of the claims. Hints that he has more than a little confidence in the observations include his explicit attribution (very unusual, for him) of these reports not just to fishermen but to experienced fishermen, ἐμπειρικοὶ ἁλιῆς. A second hint is that he adds the story about an animal of the last class having once been caught, even lending verisimilitude by including the detail about the kind of line on which it was caught (perhaps he also thinks this tells him something about its diet). In other contexts he similarly flags his confidence in observation reports, as at HA 504b26, where, apropos of the suckling behavior of dolphins, he says καὶ τοῦτο ὦπται ἤδη ὑπό τινων φανερῶς, "this has been seen clearly by some"—not only seen, but seen clearly, and by more than one authority.

At the other end of Aristotle's use of professional testimony are instances where he tells us that he is uncertain not about the source but instead about the evidence itself, regardless of what the professionals say. For example, in discussing the senses in testacea, both how many they may have and how good their perception might be, he says:

> περὶ δ' ὄψεως καὶ ἀκοῆς βέβαιον μὲν οὐδέν ἐστιν οὐδὲ λίαν φανερόν, δοκοῦσι δ' οἵ τε σωλῆνες, ἄν τις ψοφήσῃ, καταδύεσθαι, καὶ φεύγειν κατωτέρω, ὅταν

αἴσθωνται τὸ σιδήριον προσιόν . . . καὶ οἱ κτένες, ἐάν τις προσφέρῃ τὸν δάκτυλον, χάσκουσι καὶ συμμύουσιν ὡς ὁρῶντες. καὶ τοὺς νηρείτας δ' οἱ θηρεύοντες οὐ κατὰ πνεῦμα προσιόντες θηρεύουσιν, ὅταν θηρεύσωσιν αὐτοὺς εἰς τὸ δέλεαρ, οὐδὲ φθεγγόμενοι ἀλλὰ σιωπῶντες ὡς ὀσφραινομένων καὶ ἀκουόντων· ἐὰν δὲ φθέγγωνται, φασὶν ὑποφεύγειν αὐτούς.[33]

Concerning sight and hearing nothing is certain or very clear. When someone makes a noise, razor clams pull down and hide deeper as though they perceive the iron approaching . . . And scallops, if someone brings a finger near them, open and close as though they saw it.[34] Additionally, people hunting sea snails never go upwind of them once they have lured them to the bait, nor do they talk but instead keep quiet, as though the snails can smell and hear. If they talk, they say that the snails will flee.

The source of Aristotle's puzzlement in the face of this evidence is not entirely clear. Likely the apparent behavior of the animals is counterbalanced by those animals' seeming lack of the sensory organs they would need in order to see or hear. In the last of the examples, which relies on the testimony of snail hunters, we may be faced with the problem flagged earlier of the inherent reliability of the uneducated. That this is a genuine concern for Aristotle is clear from another passage where he sets out his own knowledge of the octopus as superior to what the fishermen think they know:

ἡ δὲ τῆς πλεκτάνης τοῦ ἄρρενος διὰ τοῦ αὐλοῦ δίεσις ἐπὶ τῶν πολυπόδων, ᾗ φασιν ὀχεύειν πλεκτάνῃ οἱ ἁλιεῖς, συμπλοκῆς χάριν ἐστὶν ἀλλ' οὐχ ὡς ὀργάνου χρησίμου πρὸς τὴν γένεσιν.[35]

In octopus, the insertion of the male's tentacle into the funnel [of the female], by means of which tentacle fishermen say they mate, is [really] for the sake of embracing, and not an organ useful for generation.

There are other passages where he worries about what professionals report, and a great number of passages where reports or stories from unnamed persons, professional or otherwise, are dismissed outright.[36] In one interesting passage, Aristotle pits one group of observers against another on one subject, only to tell us that they do, however, all agree on a second subject:

> ἔχει [ὁ σπόγγος] δὲ καὶ αἴσθησιν, ὡς φασίν. σημεῖον δέ· ἐὰν γὰρ μέλλοντος ἀποσπᾶν αἴσθηται, συνάγει ἑαυτὸν καὶ χαλεπὸν ἀφελεῖν ἐστιν. ταὐτὸ δὲ τοῦτο ποιεῖ καὶ ὅταν ᾖ πνεῦμα πολὺ καὶ κλύδων, πρὸς τὸ μὴ ἀποπίπτειν. εἰσὶ δέ τινες οἳ περὶ τούτου ἀμφισβητοῦσιν, ὥσπερ οἱ ἐν Τορώνῃ.[37]
>
> [The sponge] has perception, so they say. The indication [of this] is that it perceives when someone is going to harvest it, contracts, and becomes difficult to remove. It also does this whenever there is wind or rough water so that it doesn't get detached. But there are some who dispute this, such as those in Torone.

One gets the impression that Aristotle is happy to allow sponges to have some level of sensation, although he is in the end noncommittal, telling us only what he knows from the various reports. What everyone agrees on, though—and he makes the point quite forcefully—is that there is one kind of sponge that clearly does have sensation: ἔστι δ' ἄλλο γένος ὃ καλοῦσιν ἀπλυσίας διὰ τὸ μὴ δύνασθαι πλύνεσθαι· . . . ὁμολογεῖται δὲ μάλιστα παρὰ πάντων τοῦτο τὸ γένος αἴσθησιν ἔχειν καὶ πολυχρόνιον εἶναι, "there is another kind of sponge which are called 'dirty sponges' because they cannot be washed . . . It is completely agreed by everyone that this kind has perception and is long-lived."[38] Aristotle could hardly be more emphatic here. It is agreed, it is agreed completely, and it is agreed completely by everyone.

Negative Observations

The last class of observation statements in Aristotle consists of what we might call "negative observations," and these pose their own special epistemological problems. By negative observations, I refer to moments when Aristotle tells us that some behavior or bodily structure, perhaps common in other species, has never been observed in some different group of animals. As with his positive observation claims, where Aristotle commonly signals to the reader his degree of confidence in individual observations, he is likewise quick to qualify his certainty with regard to negative statements. Qualifications can range from the confident-but-still-cautious "this has never been observed," to the frank admission that "sufficient observations have not yet been made."

Even when he clearly thinks that a "never-been-seen" phenomenon implies categorical nonexistence, he is surely aware that these assertions are not, strictly speaking, logically binding. So in the HA, in discussing the generation of eels, he tells us, without any hint of doubt, that αἱ δ' ἔγχελυς οὔτ' ἐξ ὀχείας γίγνονται οὔτ' ᾠοτοκοῦσιν, οὐδ' ἐλήφθη πώποτε οὔτε θορὸν ἔχουσα οὐδεμία οὔτ' ᾠά, "eels are not generated by mating, nor do they bear eggs, and none has ever been found having any milt or eggs."[39] The conclusion that eels do not generate by mating comes across as a statement of fact, based in part on the negative observation that they have never, ever, been found with milt. Similarly, in trying to work out how bees are generated, Aristotle shows full confidence in the conclusions that (a) among worker bees there cannot be both males and females, and (b) of workers and drones, they are not one of them male and the other female. His primary proof of these points is that if either of the two were the case, then someone should have seen bees mating by now, but no one has:

> κοινὸν δὲ καὶ πρὸς τὴν ἐξ ἀλλήλων γένεσιν καὶ πρὸς τὴν ἐκ τῶν κηφήνων, καὶ χωρὶς καὶ μετ' ἀλλήλων, τὸ μηδέποτε ὤφθαι ὀχευόμενον μηθὲν αὐτῶν· εἰ δ' ἦν ἐν αὐτοῖς τὸ μὲν θῆλυ τὸ δ' ἄρρεν, πολλάκις ἂν τοῦτο συνέβαινεν.[40]
>
> The common argument against both the theory of generation by workers mating with each other and that of generation from drones (either mating with other drones or with the workers) is that none of them has ever been seen mating. And if there were both male and female among them, this would happen often.

Again, "not ever seen" (sufficient opportunity implied) seems here to be proof enough of "not ever happening." (We shall have more to say on bees presently.)

A weaker version of this sort of negative observation modifies the "not-ever" formula (οὐ πώποτε, μηδέποτε, etc.) to say instead "not yet" (οὔπω, etc.), which seems inherently to offer at least a little more in the way of an admission that our understanding may one day change if more evidence comes to light. A lovely example comes in Aristotle's discussion of whether there are any animal species that have only females and no males:

> εἰ δ' ἐστί τι γένος ὃ θῆλυ μέν ἐστιν, ἄρρεν δὲ μὴ ἔχει κεχωρισμένον, ἐνδέχεται τοῦτο ζῷον ἐξ αὑτοῦ γεννᾶν. ὅπερ ἀξιοπίστως μὲν οὐ συνῶπται μέχρι γε τοῦ

νῦν, ποιεῖ δὲ διστάζειν ἐν τῷ γένει τῷ τῶν ἰχθύων· τῶν γὰρ καλουμένων ἐρυθρίνων ἄρρην μὲν οὐθεὶς ὦπταί πω, θήλειαι δὲ καὶ κυημάτων πλήρεις.[41]

If there is a species which is [entirely] female and has no [distinct] male, it may be possible for this animal to generate from itself. This has not been reliably observed thus far, at any rate, but there is [an example] among fish that gives us pause, for in the so-called erythrinoi, no male has yet been observed, and the females are full of eggs.

Aristotle is being both admirably clear and admirably careful. He readily admits that nothing is certain, since sufficient observations have not yet been made to cinch the case. However—and here things get very interesting—he says that one species of fish in particular, the erythrinus, has been observed enough that he is willing to entertain the real possibility that it is all-female and that it reproduces parthenogenically. But he does not go so far as to say that the matter is definite. Quite the contrary, just after saying this he repeats his statement that the proof is not yet reliable (οὔπω πεῖραν ἔχομεν ἀξιόπιστον), but he is very clearly floating the erythrinus as the best contender of which he is aware for a species with only one sex. Indeed, the erythrinus is one of Aristotle's go-to species when he wants to point to animals that are an exception to the otherwise standard patterns of sex differentiation and reproduction, and elsewhere he has no problem stating the case as more or less factual: νῦν δ' οἱ μὲν ἔχουσι θορικὰ οἱ δ' ὑστέρας, καὶ ἐν ἅπασιν ἔξω δυοῖν, ἐρυθρίνου καὶ χάννης, αὕτη ἐστὶν ἡ διαφορά· οἱ μὲν γὰρ θορικὰ ἔχουσιν, οἱ δ' ὑστέρας, "now then, [in any given fish species] some have seminal parts, and others have wombs, and [this is the case] in all but two: the *erythrinus* and the *channa*, and this is the difference: the one has seminal parts and the other wombs."[42]

A similar case occurs in a discussion of another fish, the *rhinobatos,* which he thinks may be an animal hybrid:

τῶν μὲν . . . ἰχθύων παρὰ τὰς συγγενείας οὐδὲν ὦπται συνδυαζόμενον, ῥίνη δὲ δοκεῖ μόνη τοῦτο ποιεῖν καὶ βάτος· ἔστι γάρ τις ἰχθὺς ὃς καλεῖται ῥινόβατος· ἔχει γὰρ τὴν μὲν κεφαλὴν καὶ τὰ ἔμπροσθεν βάτου, τὰ δ' ὄπισθεν ῥίνης, ὡς γιγνόμενος ἐξ ἀμφοτέρων τούτων.[43]

Among fish, none has been seen mating outside its species. But the *rhinē* and the *batos* do seem to do this, for there is a fish called the

> *rhinobatos,* and it has the head and foreparts of a *batos* and the rear parts of a *rhinē,* as though it were generated from both of these.

Elsewhere he again shows this same qualified confidence: ἐπὶ δὲ τῶν θαλαττίων οὐθὲν ἀξιόλογον ἑώραται, δοκοῦσι δὲ μάλιστα οἱ ῥινοβάται καλούμενοι γίγνεσθαι ἐκ ῥίνης καὶ βάτου συνδυαζομένων, "concerning marine animals, nothing worth reporting has been observed, but the so-called *rhinobatos* very much seems to be produced by the mating of the *rhinē* with the *batos.*"[44] In both cases he signals that he is pretty sure that this fish is a hybrid, but he also clearly indicates that he is still one step removed from certainty. He has, it seems, not actually seen the two different species mating with each other, nor, one supposes, has a *rhinobatos* turned up in a pond that had previously contained only *rhinai* and *batoi.*

Difficult Observations

There are a number of interesting passages in which we seem to catch Aristotle in the middle of an as-yet uncompleted research project. In describing how cuttlefish eggs develop, for example, he tells us that once the egg is laid there appears within it something like a hailstone (a phenomenon he has observed in the development of birds' eggs, as well).[45] This hailstone slowly transforms into the young cuttlefish, but Aristotle is not sure about how its umbilical attachment works: ποία δέ τίς ἐστιν ἡ πρόσφυσις ἡ ὀμφαλώδης, οὔπω ὦπται, πλὴν ὅτι αὐξανομένου τοῦ σηπιδίου ἀεὶ ἔλαττον γίγνεται τὸ λευκόν, καὶ τέλος, ὥσπερ τὸ ὠχρὸν τοῖς ὄρνισι, τούτοις τὸ λευκὸν ἀφανίζεται, "what the umbilical outgrowth is like has not yet been observed, but only that as the little cuttlefish grows the white always diminishes and in the end the it disappears, just like the yolk does with birds."[46] Assuming that these reflect Aristotle's own observations, which I think is quite plausible in this instance, we see him here telling us both what he has managed to see, as well as what he hopes to one day catch a glimpse of. He readily admits that he has not yet managed to see the umbilical outgrowth (perhaps he has never had a specimen at just the right stage of development).

Another observation that is sorely tempting to ascribe to Aristotle himself—but which is, alas, unconfirmable—concerns the mating of wasps:

ὠμμένοι δ' εἰσὶν ὀχευόμενοι ἤδη καὶ τῶν ἄλλων τινές· εἰ δ' ἄκεντροι ἄμφω ἢ κέντρα ἔχοντες, ἢ ὁ μὲν ὁ δ' οὔ, οὔπω ὦπται. καὶ τῶν ἀγρίων ὀχευόμενοι ὠμμένοι, καὶ ὁ ἕτερος ἔχων κέντρον· περὶ θατέρου δ' οὐκ ὤφθη.[47]

Some of the other [kind] have by now been observed mating, but whether both were stingerless or had stingers, or one did and the other did not, has not yet been observed. Some wild wasps have been observed mating, and one had a stinger. But as for the other, it was not observed.

Here there are a couple of indications that the observation of wild wasps mating was a singular experience: that only one of the mating pair was observed closely enough to determine whether it had a stinger has a distinct ring of missed opportunity to it. Perhaps more telling, there is a shift in the grammatical structure: the verbs of the first clauses (those pertaining to the observation of the "other kind" of wasp) are all in the perfect tense (ὦπται, ὠμμένοι), and the point about whether one or both have stingers is worded as "[this] has not yet been observed."[48] When he turns to the wild wasps, again we see a perfect participle (they "have been observed mating," in my translation), but when he comes to describe more particularly the observation of whether the latter wasps had stingers, he switches to the aorist tense when he tells us that one of the pair was not observed, dropping a hint that this was a unique experience.[49]

Whether it was his own observation or not is impossible to determine definitively, but the passage does indicate a level of dedication on the part of the investigator, whoever that person may have been. For Aristotle says, twice, that wasps keep their stingers "on the inside,"[50] meaning he is likely limiting his class of wasps, σφῆκες, to what we now would call stinging wasps, rather than including our (distinct) class of parasitic wasps, which have their stinger-like ovipositors permanently on the outside (perhaps these fell under his third class of bee-like insects, the *anthrenae* [ἀνθρῆναι]).[51] If he is limiting "wasps" to stinging wasps, then the investigator who tried to determine whether in this particular pair of mating wasps each of them had stingers or not would have had to physically interfere with the mating pair and, in fact, alarm them enough that they would produce their stingers (perhaps in their panic at being held by fingers or tweezers, perhaps being coerced to actually sting the observer). Whatever the case, we are told, the observer was unable on this one occasion to test both wasps, and the question must have been left for another day.

This observation of the wasps' mating contrasts nicely with a passage in the *History of Animals* where Aristotle is explaining that insects in general mate with the female, pushing her "duct" (πόρος) up into the male from below:

> τοῦτο δ' ἐστὶ φανερόν, ἄν τις διαιρῇ τὰς ὀχευομένας μυίας. ἀπολύονται δ' ἀπ' ἀλλήλων μόλις· πολὺν γὰρ χρόνον ὁ συνδυασμός ἐστι τῶν τοιούτων. δῆλον δ' ἐπὶ τῶν ἐν ποσίν, οἷον μυιῶν τε καὶ κανθαρίδων.[52]
>
> This becomes clear if one separates flies while they are mating. They come apart from each other with difficulty, for the coupling of insects like these takes a long time, which is clear in everyday insects such as flies and beetles.

This experience is clearly of a different sort than the missed opportunity with the wasps. Anyone, τις, says Aristotle, can check this. The example is especially vivid because the animals from which we learn about this method of mating are around us all the time.

These claims that something remains to be seen are of a kind with other passages where he says that something has not yet been "sufficiently" (ἱκανῶς) observed.[53] Such admissions that more work needs doing are littered throughout the biological corpus, and Aristotle is clearly not shy about using them.[54] "Not yet sufficiently seen" contrasts nicely with other passages where Aristotle tells us the opposite, that a thing has been seen sufficiently to be reliable.[55]

From a philosophical point of view, perhaps the most interesting and detailed expansion of the claim of insufficient observation comes in his discussion of bees, where Aristotle seems almost to be working out his theory in real time as the text progresses, weighing and thinking through each possibility and problem before the reader's very eyes.

He begins his discussion of bees in the *Generation of Animals* with a candid statement: "The generation of bees," he says, "is a great puzzle," ἡ δὲ τῶν μελιττῶν γένεσις ἔχει πολλὴν ἀπορίαν.[56] He then proceeds to think through all the many possibilities that may apply and outlines the pros and cons of each. The resulting list of possibilities, although foreign to the modern eye, is admirably well thought through. Perhaps, he begins, bees collect their young from plants and flowers, the larvae being either spontaneously generated or else laid there by some other insect.[57] Perhaps bees generate the young themselves (but then he has to weigh the evidence for how this would work when there is, uniquely, a threefold division within the species—kings,[58] workers, and drones—instead of the usual two sexes). Perhaps the drones are generated by other insects, or generated spontaneously somewhere outside the hive, and the workers collect them from wherever they are found, while the kings and workers mate to reproduce their own kinds (a theory he says "some" hold). Perhaps the kings (or the workers, or the drones) alone reproduce themselves without mating, and the other two kinds mate with each other. Perhaps each kind reproduces its own somehow. For each of these last three theories, where some or all of the kinds reproduce their own, he then needs to decide whether this is by mating with another of the three kinds, mating with their own kind, or without mating at all.

Against the collecting-from-elsewhere theories, he argues that (a) this would mean we should find bee larvae not-yet-collected and out in the wild, which we don't, and (b) if these larvae were left there by some other animal, they would need to turn into that animal rather than into bees; and relatedly, (c) he cites a law that no animal cares for offspring that does not appear to be οἰκεῖον, "proper," to it.[59] Against workers and drones mating with each other, he finds himself unable to see how one of them could be male and the other female: workers can't be female and drones male because the workers have stingers, and no other species of animal has "defensive weapons" assigned only to its females. But the workers can't be male either, because they care for their young and—we're about to learn something about ancient Greek parenting styles—no males make a habit of caring for their young. Against the theory that one or the other classes of bee mates with its own kind, he points out that there does not seem to be any sexual difference between worker and worker that would make this possible and also that workers do not seem to be produced unless the kings are in the hive (but drones, puzzlingly, do seem to be produced in the absence of kings, he says).[60]

He continues in this vein, deliberating over each of the multiple possibilities in turn, for several pages before settling on what he sees as the likeliest scenario given the evidence: because drones can be produced even when neither the king nor other drones are present, drones must be produced by workers, without copulation, and this means that workers must have both male and female within themselves.[61] (He says that this manner of reproduction "seems to correspond with" [φαίνεται συμβαῖνον] "certain fish," later naming the *erythrinus* and *channa* but also pointing out that, unlike worker bees producing drones, even these odd fish do not generate something different from themselves, so the parallel is not perfect.)[62] Finally, he concludes that workers themselves must be produced asexually by the kings, who also produce their own kind from time to time in the same way. As strange and unparalleled as this method of populating a species may be, Aristotle finds some solace in his discovery of a mathematical series at play: kings generate two things (kings and workers), workers generate one (only drones), and drones generate nothing.[63]

At this point Aristotle may, if we are not careful, be read as being more confident than he really is. But no—he is about to tell us that there is still non-negligible uncertainty surrounding bees. Little hints in this direction are easy to have missed along the way: for both of his major conclusions he introduces them as "the remaining (possibility)," λείπεται, rather than asserting their certainty more definitively.[64] Toward the end of the discussion he throws in a little string of additional facts that, rather than "proving" the case, simply "agree with it," ὁμολογούμενον δ᾽ ἐστί.[65] But it is when he comes to wrap up the section on bees that Aristotle becomes much more candid and reflective. He ends his discussion with one of the more remarkable methodological pronouncements in ancient science:

> ἐκ μὲν οὖν τοῦ λόγου τὰ περὶ τὴν γένεσιν τῶν μελιττῶν τοῦτον ἔχειν φαίνεται τὸν τρόπον, καὶ ἐκ τῶν συμβαίνειν δοκούντων περὶ αὐτάς· οὐ μὴν εἴληπταί γε τὰ συμβαίνοντα ἱκανῶς, ἀλλ᾽ ἐάν ποτε ληφθῇ τότε τῇ αἰσθήσει μᾶλλον τῶν λόγων πιστευτέον, καὶ τοῖς λόγοις, ἐὰν ὁμολογούμενα δεικνύωσι τοῖς φαινομένοις.[66]

This, then, seems to be the state of affairs concerning the generation of bees, based on our reasoning and on the facts as they appear with respect to bees. But the facts have not been sufficiently established. If they ever are, then observation must be given more credence than reasoning,

> although reasoning, too, can be given credit if it be shown to agree with the phenomena.

Now, "agreeing with the phenomena" often takes a broader meaning in Aristotle than just "agreeing with perception,"[67] but here Aristotle makes the primacy of sensory evidence over reason explicitly and abundantly clear: he says that it is αἴσθησις, perception, that is to be primary, and reasoning is to be subject to the constraints of that evidence. We can see, in fact, that this is a very good description of Aristotle's methodology in the biological books, where he seems always attentive to just where the evidence is pushing him, and he is quite happy to list exceptions to any of his general rules, should such exceptions appear.

That Aristotle is careful and deliberate about how he reports his certainty in the various types of observation claim seems at this point clear. More importantly, we see that he often signals his confidence in observation claims on a case-by-case basis, indicating to his reader not just how reliable he thinks testimony is in general, but how reliable he thinks each particular instance of testimony—this animal, seen by this person—might be. And of course testimony is only part of the story. I remarked at the start of this chapter that Aristotle is widely thought to have performed a good number of his own observations, most strikingly in the study of animal anatomy, and his reports of observations, whether explicitly his own or simply unattributed, are clearly marked by Aristotle to signal the degree of trust he has in them. Whether he states them as bare facts about the world or simply marks them as "what is seen," or qualifies them with rhetorical formulae to signal his moderated certainty, Aristotle is very careful to maintain as much clarity as possible with regard to his own certainties and uncertainties, his own questions and puzzlements, and his own trusts and suspicions about individual observations for his reader.

A Blossoming of Creatures

ALTHOUGH ARISTOTLE is not the only author to talk about spontaneous generation in antiquity, he is both the most thorough and the most influential. Later spontaneous generation theories are, until a very late date, often notably Aristotelian in character. Having said that, by the time spontaneous generation got into the hands of medieval European scholars significant changes had already taken place.

Partly this was owing to later ancient and Islamic authors taking on board other ideas that were circulating here and there in the literature and in folklore; but it was also partly owing to various classical authors after Aristotle rethinking the evidence, the mechanisms, and the possibilities for what comes to be and how that process happens.

One prominent and widespread claim, particularly in Roman literary sources, is what is known among classicists as the *bougonia*, the idea that the rotting of a cow's corpse generates bees.[1] Varro, an early source for this claim, cites the authority of a certain Archelaus, a source he uses multiple times in his *Rerum rusticarum*. Varro gives us a pair of Greek verses whose contents will deeply inform later tradition: βοὸς φθιμένης πεπλανημένα τέκνα and ἵππων μὲν σφῆκες γενεά, μόσχων δὲ μέλισσαι, "[bees] are the offspring of rotting cattle" and "wasps come from horses, bees from calves."[2] This pairing of bees with cattle and wasps with horses becomes quite cemented and, along with the cognate claim that beetles come from asses, will do important work in the tradition, as we shall see, standing as an important set of examples for how spontaneous generation is law-like rather than random. Randomness, after all, is antithetical to the rationally ordered cosmos that almost all of the actors in this story believed themselves to occupy.

In many sources, we find a peculiar (and peculiarly expensive) method for preparing the carcass so that bees will be produced. By far the most influential and widely discussed version of this technique is given by Vergil, in a stunning passage that alternates quite jarringly between languishing pastoral tranquility and visceral butcher-shop slaughter:

> exiguus primum atque ipsos contractus in usus
> eligitur locus; hunc angustique imbrice tecti
> parietibusque premunt artis, et quattuor addunt,
> quattuor a ventis obliqua luce fenestras.
> tum vitulus bima curvans iam cornua fronte
> quaeritur; huic geminae nares et spiritus oris
> multa reluctanti obstruitur, plagisque perempto
> tunsa per integram solvuntur viscera pellem.
> sic positum in clauso linquunt, et ramea costis
> subiciunt fragmenta, thymum casiasque recentis.
> hoc geritur Zephyris primum impellentibus undas,
> ante novis rubeant quam prata coloribus, ante
> garrula quam tignis nidum suspendat hirundo.
> interea teneris tepefactus in ossibus umor
> aestuat, et visenda modis animalia miris,
> trunca pedum primo, mox et stridentia pennis,
> miscentur, tenuemque magis magis aera carpunt,
> donec ut aestivis effusus nubibus imber
> erupere, aut ut nervo pulsante sagittae,
> prima leves ineunt si quando proelia Parthi.
> quis deus hanc, Musae, quis nobis extudit artem?
> unde nova ingressus hominum experientia cepit?[3]

> First a small narrow place is chosen for the purpose. They close it in with low walls and narrow tiles covering, and they add four windows with slanting light facing the four winds. They seek a young bull, shooting forth two-year-old horns from his head.[4] Although he struggles mightily, they plug his two nostrils and the breath of his mouth. Killing [him] with blows, the pounding dissolves his viscera through the intact skin. So lying in this cell, they put under his ribs bits of branches and fresh thyme and cassia. This is done when the west wind first stirs the

> waves, before the meadows shine with their newfound colors, before the chirping swallow hangs her nest in the beams. During that time the moisture ferments, having been warmed in the softened bones. And living things, visible to wondering eyes, swarm; first without feet, soon with buzzing wings. More and more they take to the light air until they burst forth like rain pouring from summer clouds or like arrows from their singing string when the swift Parthians first enter a battle.
>
> What god, O Muses, forged for us this trick? Where did this practice, a strange thing among men, get its origin?

Giambattista Della Porta, in his *Magia naturalis* of 1589, quotes the passage almost in full, calling it "most elegantly written."[5] Not so concerned with matters of Latin style, the 1658 English translation of Della Porta casts a moral eye on the practice, adding that this is "a cruel manner of making bees in a house, but it is a very ready way." Vergil, just before the passage quoted above, lists all the countries where this manner of generating bees is practiced (all of them foreign, as it turns out) and calls the technique their "salvation." Similarly, Della Porta goes on to tell us that the best kinds of bee are brought forth from a young ox, although poorer quality bees can be generated from the flesh of "a cheaper animal," *vilioris-que carnibus animalis*.[6] This nod to economy serves to remind us of an undercurrent that runs though many of the recipes we find for spontaneously generating bees and other animals, which is that such animals have financial value—spontaneous generation as an economic activity.

Bees and wasps, then, come to be added to the class of animals that are generated spontaneously in the period between Aristotle and Albertus Magnus, but these are not the only ones. Depending on the author, any number of creatures—up to and including humans—were thought to be so born.[7] Thus Ibn Tufail accounts for the completely isolated island existence of his character Hayy ibn Yaqdan in the twelfth-century novel of the same name: when the perfectly tempered light that streams down off the coast of India fell on clay in which just the right balance of hot, cold, wet, and dry had been achieved "over the years,"[8] the hero of the novel sprang into existence independently of any parents or, indeed, of any society. The framework of this story is rooted in earlier accounts of spontaneous generation going back to Avicenna, who had argued that animals and people were spontaneously generated after natural disasters such as the regular, if rare, floods that he thought punctuated history. In the damp mud after the flood, when just the right mixtures happened to occur, up sprang all the plants and animals we see now.[9] Rooted in an idiosyncratic ontology and reliant on a semidivine entity called the Giver of Forms, Avicenna's claim that humans could be spontaneously generated did not gain a great deal of traction in the later Islamic and Latin traditions, but it did circulate widely enough to be the subject of a Paris condemnation in 1277.[10] The generation of humans also functioned as a kind of limiting case for more widespread discussions of whether perfect animals could be spontaneously generated or whether this means of coming-to-be was limited to imperfect animals (we return to this topic in chapter 4).[11]

Lucretius, too, had earlier argued that humans could be spontaneously generated—more to the point, that they had been, once upon a time, but were no longer. The idea is rooted in his reading of Epicurean atomism, where the world is made up of only two things: atoms and the void they move in.[12] Everything, including the souls of animals, plants, and humans, is made of atoms, blazing around in a perfect vacuum at inconceivable speeds. The gods, too, are made entirely of atoms. And here is where Epicureanism marks a radical break from other ancient philosophical traditions: Lucretius insists that the gods do not interfere in our world. They are not moved by our prayers, they do not care if we perform sacrifices to them, they do not get angry at our transgressions, and they do not watch over us—or indeed watch us at all. Any concern with us, positive or negative, that people imagine the gods might have with the silly goings-on down here on earth or anywhere else in the cosmos would be entirely beneath their dignity. Instead, they exist outside of our cosmos, in a state of perfect carefree-ness (which for an Epicurean means a state of perfect happiness) and so serve as ethical role models for us to contemplate and emulate. This position of complete noninterference in the cosmos is an unusual one in ancient philosophy and would lead many readers of Lucretius—from antiquity through the Renaissance and beyond—to suspect that he and his followers were really just disguising a bald atheism by paying lip service to the existence of divinity.

Since everything in Epicureanism was explained by the motions of atoms in a void, every physical phenomenon, every emotion, every sensation, and the growth, aging, and death of every animal needed to be reducible to the impacts of hard, indivisible atoms, having the properties only of extension, shape, and weight.[13] Epicurean cosmology argued that the universe had no first beginning, no creation, but that atoms have been colliding and intertwining, creating planets, animals, and even universes, for all time. Our own earth came to be at some time in the distant past, and through a natural process of atomic interactions, began at some point to give birth to living beings:

> tum tibi terra dedit primum mortalia saecla.
> multus enim calor atque umor superabat in arvis.
> hoc ubi quaeque loci regio opportuna dabatur,
> crescebant uteri terram radicibus apti.[14]

> At that time, you see, earth first delivered up mortal breeds. Indeed, a great heat and moisture abounded in the fields. Then, when some suitable parcel of land was furnished, wombs grew, seizing the earth with their roots.

In a nod to Empedocles,[15] Lucretius also tells us that once upon a time nature give birth to many of what we might call "experimental" forms, which were, in the end, not viable for one reason or another:

> multaque tum tellus etiam portenta creare
> conatast mira facie membrisque coorta,
> androgynem, interutrasque nec utrum, utrimque remotum,
> orba pedum partim, manuum viduata vicissim,
> muta sine ore etiam, sine vultu caeca reperta.[16]

> And at that time the earth had tried to produce many monstrosities, arising with wondrous face and limbs: a manwoman—between both but neither, far from either—[some animals] lacking feet, or even without hands, silent without mouths, or found blind, without a face.

A process of natural selection weeded these out, either through their susceptibility to predation or by starvation.

To feed its fledgling animals (the viable ones, at least) the earth in those early days produced a "nectar, like unto milk,"[17] *sucum . . . consimilem lactis*. In fact, this whole long passage is rife with metaphors of the earth as a mother to these first animals, but unfortunately that is about as far as Lucretius goes in explaining the actual mechanism of spontaneous generation. He never tells us precisely how the earth produced this life, although there are hints here and in an earlier book. In the present passage, he makes a standard, if vague, nod to heat and moisture as preexisting conditions, and he adds that this is also the same mechanism that produces maggots and other animals to this day:

> multaque nunc etiam existunt animalia terris,
> imbribus et calido solis concreta vapore;
> quo minus est mirum si tum sunt plura coorta
> et maiora, nova tellure atque aethere adulta.[18]

> For even now many living things come forth from the earth, coalesced by the rains and the hot exhalation of the sun, and so it is not surprising if at that [earlier] time many and larger ones had come forth and grown when the earth and the air were [still] young.

But if spontaneous generation was the mechanism through which all currently existing species of animals and plants, including humans, came into existence in the remote past (they could hardly have "fallen from the sky," *neque de caelo cecidisse animalia possunt*, he adds),[19] why then do we not still see the earth giving birth to lions and people today? The answer, for Lucretius, brings us back to his metaphor of the earth as mother: *sed quia finem aliquam pariendi debet habere, / destitit, ut mulier spatio defessa vetusto*, "because she must have some end to her parturition, she has become barren, like a woman tired with old age."[20] Notice here the subtly amusing dig at Aristotelian theories of generation: the only end—he uses the word *finis*, which in an Aristotelian context it would be natural to translate as "final cause"—the only *finis* the generative capacity of the earth has (or better, needs) is that which is necessitated by the impartial and inexorable march of time.[21] Indeed, even the structure of the line itself seems to playfully foist exactly this misreading on the unsuspecting non-Epicurean: as a self-contained unit, *sed quia finem aliquam pariendi debet habere* might at first blush be read by many a philosopher as "but because there must be some final cause of generation."[22] Then, with the very next word on the very next line, Lucretius lets this reader have it: *destitit*, "she has become barren"—no final cause, just the other, more prosaic kind of finish. This is Lucretius putting his finger on one of the most important vulnerabilities in the Aristotelian theory of spontaneous generation: its difficulty specifying the final cause of spontaneously generated organisms. For Lucretius, one of the strengths of the Epicurean system over virtually all of its competitors is that it can explain everything in terms of billiard-ball interactions between atoms, without recourse to such final causes.

Returning now to the mechanism of spontaneous generation for Lucretius, apart from mentioning the need for heat and moisture, there is one section where he goes into slightly more detail that will be worth looking at. He begins by asking what prevents his reader from believing that something endowed with sensation can come to be from mere atoms, which have no sensation whatsoever.[23] Do we imagine that if we mixed together raw matter, sticks and stones and muck, we should get nothing but a lifeless mess? No—everything depends for Lucretius on the size, shape, and arrangement of the atoms in that mess. If the finest atoms come into just the right relation with other atoms, if they become, as he repeatedly calls it, "yoked together," then sense-possessing life can arise. The finest atoms become a soul, yoked to the coarser atoms of the body, and the two work together until death, when the soul atoms lose their hold on each other and on the atoms of the body, and all go their separate ways toward whatever future encounters with other atoms may befall them:

> quarum nil rerum in lignis glaebisque videmus;
> et tamen haec, cum sunt quasi putrefacta per imbris,
> vermiculos pariunt, quia corpora materiai
> antiquis ex ordinibus permota nova re
> conciliantur ita ut debent animalia gigni.[24]

> We cannot see these processes [happen] in the sticks and soil. And yet, when they become almost putrefied by the rain, they give birth to little worms because the atoms of the material, stirred up from their former configurations, are brought together in a new way, such that organisms will come forth.

We don't get a lot in the way of specifics, but we can at least see that he is pushing the idea that new life is nothing more than just the right combination of atoms,[25] happening under the influence of moisture and heat. With this explanation, there need be no soul-containing preexistent matter, no pneuma. Life, sensation—these are nothing more than by-products of particular combinations of atoms, which happen to be in moist soil, coming together through the action of heat and putrefaction.

Lucretius never mentions which particular species he believes can still be spontaneously generated by today's aged earth, apart from a couple of references to worms (arising from "stinking excrement," *stercore de taetro*, and from "corpses exhaling [them] from viscera now grown putrid," *cadavera rancenti iam viscere vermes/expirant)*, as well as one mention that "the earth, exhausted,[26] now struggles to bear even small organisms,"[27] *effetaque tellus/vix animalia parva creat.*

What stands out in Lucretius is the idea that life can come into existence, unproblematically, from nonlife. Recall that Aristotle had mixed pneuma, and therefore soul-heat, into all matter, which may not be quite the same thing as putting soul itself into everything (Aristotle hedges his wording slightly: τρόπον τινὰ πάντα ψυχῆς εἶναι πλήρη, "in a certain sense, everything is full of soul"),[28] but in its context it reads to me as if he must mean that everything really is full of at least soul-in-potential. So, too, the power of souls is said by Aristotle to have a share in "a more divine body than the four elements," σώματος ἔοικε κεκοινωνηκέναι καὶ θειοτέρου τῶν καλουμένων στοιχείων.[29] Now one may be troubled by the word ἔοικε in that passage: the power of souls "seems to share," and I am not sure to what extent Aristotle was genuinely backing away from the more positive claim that it does share, since it looks very much as if his theory of spontaneous generation is committed to the latter, stronger reading. Unfortunately, he does not elaborate. But for what it is worth many of Aristotle's later readers take him to mean the stronger version. In any case, we see in Aristotle gestures, if not outright claims, to the idea that life requires the presence of something beyond the four elements of the sublunary world, and that this extra substance, pneuma, has some sort of divine quality. Lucretius, by contrast, floats an entirely materialistic theory, which claims that life is nothing more than the chance combination of atoms.

By late antiquity, as Latin philosophy and science came under the umbrella of Christianity, neither of these theories would do. Atomism was simply too theologically pernicious and left no room for the purposive hand of

God. Aristotle was more tractable, but still there were issues around reconciling the idea that life is constantly being generated both in the earth and by sexually generating animals and plants, with the idea that God had created all life in the beginning. The most influential voice to try and integrate these two positions was Augustine of Hippo (AD 354–430).[30]

Augustine's solution was not primarily aimed at theorizing or understanding the mechanisms of generation, spontaneous or otherwise, but was instead motivated by his desire to reconcile the two different accounts of creation offered in the book of Genesis, the one at Genesis 1.1–2.4 and that at Genesis 2.5–25, as well as integrating the conflicting timelines of each version with the other. Part of the problem was that, for all that Augustine wants to read Genesis literally, he had become convinced (for reasons outlined in book 4 of his *On Genesis according to the Letter* [*De Genesi ad litteram*]), that the seven days of creation were really one.[31] The argument is complex, but he says that, although in a certain sense we can talk of God's creation of the heavens and earth and of living things as having taken place over time (the six days of creation), with new things coming into being in a particular order, at a more basic level there is really only the one simultaneous act of creation. Much of his argument hinges on what we take the "mornings" and "evenings" and "days" to mean in the first creation account in Genesis, and he concludes that

> sicut isto die paulo superius commemorato significavit unum diem factum a Deo et tunc Deum fecisse caelum et terram, cum factus est dies, ut, quomodo possemus, cogitaremus simul omnia Deum fecisse, quamvis superior sex dierum enumeratio velut temporum intervalla ostendisse videretur.[32]

> As with the word *day* mentioned just above, [here] it mean[s] the one day made by God, and that God made the heavens and the earth at that time, when the day was made. Thus, inasmuch as we can, we should consider that God made everything simultaneously, even if the earlier counting of the six days appeared to indicate intervals of time.

His commitment to a literal simultaneity runs him into trouble, however, when he comes to Genesis 2.5, where the biblical story describes a time at which there were as yet no trees or shrubs on the already-existing earth. Suddenly, by the simple expedient of springs welling up to water the land, the earth produces plants where previously there had been none. In order to avoid the claim that God must have used the moment of the bubbling springs to create all the plants—which would imply that it happened at a distinct point in time *after* he had already created the earth—Augustine posits that what happened when the land received its first water was that the rational principles of life that had been created by God in the earth at the very beginning were simply activated by the onrush of moisture.[33] This happens *secundum intervalla temporum ex illa prima conditione creaturarum, ubi facta sunt omnia simul*, "according to the passing of time after that first establishment of creatures, when everything was made simultaneously."[34] Specifically, he says in this passage that what God created in the beginning were the *primordia seminum*, the "first-beginnings of seeds."[35] His terminology for these first-beginnings varies over the course of the *De Genesi*, where he uses phrases including *causales rationes* ("causal reasons"), *rationes primordiales* ("primordial reasons"), *primordia causarum* ("first-beginnings of causes"), and the most commonly cited phrasing in the modern literature on Augustine's influential theory of generation, *rationes seminales* ("seminal reasons").

The word *reason* in all of these phrases marks a slightly tricky point of translation from the Latin: *ratio* is a word with a broad semantic field. It can mean (a) a calculation or a reckoning, (b) the faculty that calculates and reckons (reason in the sense of rationality), or (c) the more prosaic *reason*, as in the reason or explanation for why something happens, among other things. I (in the company of most modern and medieval commentators) take Augustine to be here using *ratio* in one of its other common and related meanings: referring to a guiding principle or a plan (either a plan to act or an architectural plan). Keeping this in mind, although most scholars translate Augustine's *rationes seminales* as something like "seminal reasons," I will instead use "seminal principles" throughout to avoid any potential confusion of "reasons" with "justifications."

Just as these seminal principles can explain the sudden onrush of plants upon a formerly barren-looking earth, so can they explain the spontaneous generation of animals today:

> nonnulla etiam de quibusdam minutissimis animalibus quaestio est, utrum in primis rerum conditionibus creata sint, an ex consequentibus rerum mortalium corruptionibus. nam pleraque eorum aut de vivorum corporum vitiis, vel purgamentis vel exhalationibus aut cadaverum tabe gignuntur, quaedam etiam de corruptione lignorum et herbarum, quaedam de corruptionibus fructuum: quorum omnium non possumus recte dicere Deum non esse creatorem. inest enim omnibus quoddam naturae sui generis decus, ita ut in his maior sit admiratio bene considerantis et laus uberior omnipotentis artificis, qui omnia in sapientia fecit.[36]
>
> And there is a not-inconsequential question concerning certain very tiny animals: whether they were created in the initial establishment of things or whether they come from the later corruptions of mortal things. For a great number of them are born either from the defects of living bodies (either the excrements or exhalations) or else from the wasting of dead bodies. Some come from the decay of wood and plants, and others from the decay of fruits. We cannot rightly say of all these that God is not the creator. For there is in all of them the particular beauty of the nature of their species, such that from them, when considered carefully, there should be a greater admiration and fuller praise for their omnipotent creator, who made all with his wisdom.

As for the particular mechanism, Augustine explains it this way:

> inerat iam omnibus animatis corporibus vis quaedam naturalis et quasi praeseminata et quodammodo liciata primordia futurorum animalium, quae de corruptionibus talium corporum pro suo quaeque genere ac differentiis erant exortura per administrationem ineffabilem omnia movente incommutabiliter creatore.[37]
>
> Since [the creation] there has existed, in all bodies that have lived, a certain natural power, as well as the first-beginnings of animals-to-be, "pre-sown" as it were, and sort of "woven in" [with them], which [first-beginnings] are to emerge according to their own particular species and

> type, by means of the decay of such bodies, the creator moving them all unmovingly through his ineffable direction.

So these primordia, as seminal principles, are somehow scattered throughout or woven into the fabric of the cosmos. Augustine never quite tells us if they are to be thought of as invisible bubbles, or whether they are somehow interpermeating all matter. The important thing for him—really the only important thing for him—is that they were co-created with the rest of the universe, and they simply wait around in other matter until the conditions are just right to bring them into actuality. God, even now, underwrites this process in a certain way ("moving them unmovingly") but is not in this process creating anything new in an absolute sense.

This same principle also explains sexual generation, with the only difference being that there the seminal principles, rather than interpermeating all matter, seem to nest inside of each other:

> sed etiam ista secum gerunt tamquam iterum se ipsa invisibiliter in occulta quadam vi generandi, quam extraxerunt de illis primordiis causarum suarum, in quibus creato mundo, cum factus est dies, antequam in manifestam speciem sui generis exorerentur, inserta sunt.[38]

> These [plants and animals and humans] carry "themselves again," so to speak, in themselves, invisibly, in a kind of hidden force for generating, which they draw forward from those first-beginnings of their own causes, in which they were inserted when the world was created, when the day was made, until they should come forth in the visible species of their kind.

Each sexually generating plant and animal "draws a hidden force for reproduction forward" from the principles that caused themselves, as a plant or animal, to come to be. These plants and animals had in turn been inserted in their own causes, which is to say, inserted in the primordia of their own ancestors. On this reading, the *ratio seminalis* of each successive generation preexists in the *ratio* of its parent before it, which preexists in the *ratio* of its own parent, in turn. Stacked like this, we can now make sense of the phrase *vis . . . quam extraxerunt de illis primordiis causarum suarum*, "the force . . . which they draw forward from those first-beginnings of their own causes." The individual now is carrying all of its successive generations, all of which were likewise encased (along with the current individual) in the ancestors of all of them as a whole. If we run this through time, as though we were opening successive Russian dolls, we can imagine stopping partway through, at the current individual—this animal draws forward within itself, which is to say into the future, the hidden principles of future generations (the dolls nested within) from the principles (dolls) that formerly carried the current individual, itself nesting all of its own future generations, likewise enclosed: *insunt autem illis [primordiis seminum] efficacissimi numeri trahentes secum sequaces potentias ex illis perfectis operibus Dei a quibus in die septimo requievit*, "there inhere in these [first-beginnings of seeds] the most potent numbers, which carry in themselves the potencies coming after [them], through those completed actions of God from which he rested on the seventh day."[39]

Indeed, Augustine compares spontaneous generation to sexual generation in an interesting passage in the *De trinitate*:

> ista quippe originaliter ac primordialiter in quadam textura elementorum cuncta iam creata sunt sed acceptis opportunitatibus prodeunt. nam sicut matres gravidae sunt fetibus, sic ipse mundus gravidus est causis nascentium.[40]

> Certainly, these things originally and in the beginning are all already created in the whole "weave" of the elements, so to speak, and when there is opportunity they grow forth. For just as mothers are pregnant with their young, so is the world itself pregnant with the causes of generated things.

And so all life, sexual or spontaneous, is thus protogenic: created—once and only once—in the beginning,[41] even if each thing's actual moment of generation must await suitable conditions for its actuation. Sexually generated life was created with the seminal principles of each successive generation nested within each previous generation, reiteratively, and spontaneously generated life comes from seminal principles scattered about in living matter. It is worth taking special notice of these three points (protogenesis, as well as the nesting and the scattering of seminal principles), since together they are going to play major roles in Renaissance and early modern debates about spontaneous generation, and in particular they are going to motivate the famous set of experiments with which I end this book.

Briefly, before we move on, I want to highlight one other feature of Augustine's account that I think is significant, and this is the palpable sense of awe he broadcasts when he looks at how this system is constituted. "Moving unmovingly," "ineffable direction," "particular beauty," "greater admiration and fuller praise"—Augustine clearly finds beauty and wonder in all of God's creatures, as he tells us in one particularly moving passage. Picking up on his discussion of the spontaneous generation of "the smallest animals" that we saw earlier, he continues thus:

> creat [Deus illa minutissima animalia] minima corpore, acuta sensu animantia, ut maiore attentione stupeamus agilitatem muscae volantis quam magnitudinem iumenti gradientis ampliusque miremur opera formicularum quam onera camelorum.[42]
>
> [God] creates [these tiniest animals] minute in body, living [and] keen of sense, such that with careful attention we are rendered speechless at the nimbleness of a flying mosquito more than we are at the size of a packhorse, and we are more amazed at the works of ants than at the loads [carried by] camels.

This sense of wonder recalls that of Aristotle in a similar connection, where he warns the reader not to shy away from even the most disgusting aspects of biological study, to neither cringe at the study of the parts of animal and human bodies, no matter how shameful or revolting they may seem at first, nor to turn away from the least appealing of animals:

> δεῖ μὴ δυσχεραίνειν παιδικῶς τὴν περὶ τῶν ἀτιμοτέρων ζῴων ἐπίσκεψιν. ἐν πᾶσι γὰρ τοῖς φυσικοῖς ἔνεστί τι θαυμαστόν· καὶ καθάπερ Ἡράκλειτος λέγεται πρὸς τοὺς ξένους εἰπεῖν τοὺς βουλομένους ἐντυχεῖν αὐτῷ, οἳ ἐπειδὴ προσιόντες εἶδον αὐτὸν θερόμενον πρὸς τῷ ἰπνῷ ἔστησαν, ἐκέλευε γὰρ αὐτοὺς εἰσιέναι θαρροῦντας· εἶναι γὰρ καὶ ἐνταῦθα θεούς,[43] οὕτω καὶ πρὸς τὴν ζήτησιν περὶ ἑκάστου τῶν ζῴων προσιέναι δεῖ μὴ δυσωπούμενον ὡς ἐν

> ἅπασιν ὄντος τινὸς φυσικοῦ καὶ καλοῦ. τὸ γὰρ μὴ τυχόντως ἀλλ' ἕνεκά τινος ἐν τοῖς τῆς φύσεως ἔργοις ἐστὶ καὶ μάλιστα· οὗ δ' ἕνεκα συνέστηκεν ἢ γέγονε τέλους, τὴν τοῦ καλοῦ χώραν εἴληφεν.[44]
>
> The study of the ignoble animals should not childishly cause disgust. For a great wonder inheres in all natural things, just as Heraclitus is supposed to have said to some strangers: when they were entering they saw him warming himself at the fire and they hesitated. He bid them come in without hesitation, "for even here there are gods." Thus we should approach the study of each animal without aversion, since in all of them there is something natural and beautiful. For what is purposive and nonrandom are to be found in the works of nature especially. The final cause[45] has established and given birth to its ends, and it is understood as the proper place of the beautiful.

It is a frequent refrain of authors writing about spontaneous generation, right through to its eventual death in the nineteenth century, that these insignificant creatures that come to be from putrefaction or other processes are, in their own ways, spectacular and wondrous.

Finally, we should note the range and applicability of spontaneous generation in Augustine. We have already seen him say that "the smallest animals" come into existence this way, and his account of Genesis seems to imply that, at least for the first generation of plants, all trees and herbs came to be when their seminal principles became activated on the occasion of their first watering by the spring that flooded the earth. So, too, in the *De trinitate*, Augustine seems to indicate that the first generation of all animals happened in this same way.[46] But there are other living things, new to our discussion, that Augustine sees as coming into existence through the activation of seminal principles in matter. These are the frogs and serpents and so on that are "created" by magicians and holy men, as when Moses, Aaron, and the Egyptian magicians threw down their staffs and turned them into serpents in Exodus 7. It happens by means of the same mechanism, he says, as that which causes animals to be spontaneously generated:

> nec sane creatores illi mali angeli dicendi sunt quia per illos magi resistentes famulo Dei ranas et serpentes fecerunt; non enim eas ipsi creaver-

unt. omnium quippe rerum quae corporaliter visibiliterque nascuntur occulta quaedam semina in istis corporeis mundi huius elementis latent . . . illa . . . semina unde iubente creatore produxit aqua prima natatilia et volatilia, terra autem prima germina et prima sui generis animalia . . . invisibilium enim seminum creator ipse creator est omnium rerum.[47]

Nor are those demons said to be true creators when with their help the [Egyptian] magicians, opposing the servant of God, brought forth frogs and snakes. For they did not create them. Obviously there are certain hidden seeds of all things that are bodily and visibly born, lying hidden in those elemental bodies of this world . . . Those [are the] seeds from which, at the command of the creator, the water produced the first swimming and flying creatures, and the earth the first seeds and the first animals according to their kind . . . The creator of these invisible seeds is the creator, himself, of all things.

The staffs of the magicians had contained within them the hidden seeds, the seminal principles of snakes, which, like poppy seeds or grains of corn, simply wait, dormant, for the proper conditions to be brought about, and then they unfold as living things in their own right. The extraordinary nature of the staffs-to-snakes trick may have required the aid of demons, but the physical principles those demons were manipulating in the material of the staffs and on behalf of the magicians were the same principles as those that bring about the more mundane and common spontaneous generation of animals that we see on a day-to-day basis.

With this account, and unlike Lucretius before him, what Augustine is offering is neither a critique nor an explicit alternative to Aristotle on the subject of spontaneous generation in particular—it seems as though it should be entirely possible to reconcile Aristotelian "pneuma" with Augustinian *rationes seminales* without committing undue violence to either (indeed, the Renaissance Aristotelian Fortunio Liceti makes something very like this unification). The main point of difference with Aristotle, as Augustine would see it, has to do with the question of creationism versus the "always-having-been-ness" of the world. Nevertheless, the mechanics of spontaneous generation are not quite the same in Aristotle and Augustine, and one gets the distinct impression that Aristotle's pneuma is more of a diffuse potential for life spread throughout matter, where a mass of matter can turn into either single or numberless individuals, whereas Augustine's seminal principles appear to have a one-to-one correspondence with future living creatures, in the way that a single seed generates a single shoot. Again, this is an important distinction to remember for later in my discussion, even if it is not one that Augustine himself seems to have emphasized or even to have particularly noticed.

CHAPTER FOUR

Inheritance and Innovation

THE EXPANSION OF SPECIES that we saw in chapter 3, all the way up to and including humans for some authors, would continue to be rethought and modified as theories and experiences changed coming into the European Middle Ages and Renaissance. The European Middle Ages also inherited another change to Aristotelian ideas about spontaneous generation, which was a new theory about how the form of the spontaneously generated creature comes into actuality in foamy matter: it now does so under the influence of the sun, stars, and planets. This is not just the heat of the season warmed by the sun, as it had been for Aristotle. Instead, now the form is imposed under something much more like an astrological influence.

Problems with the Aristotelian account had been noticed by a number of late ancient commentators, and the Arabic philosophical tradition found ways of dealing with and responding to them. For Avicenna, the solution lay in positing a quasi-divine being called the Giver of Forms, but this entity was entirely rejected by Aristotle's most influential Arabic reader, Averroes, and it is Averroes who passes on to the Latin-speaking tradition the new active cause of enformation for spontaneous generation: the astrological influences of the stars and planets.

It is not perfectly clear whether Averroes recognized that his importation of the sun and stars into the explanatory chain was a novel reading of Aristotle, but I suspect that he did not. Averroes was explicitly responding to a critique leveled by Themistius in the fourth century against Aristotle's theory of form. Themistius had pointed out that Aristotle's general theory of change as outlined in the *Metaphysics* was apparently incompatible with the theory of how matter changed into spontaneously generated animals. Recall that change for Aristotle is always the coming-to-be of a new form in something that previously had that form in potential but not in actuality. A carpenter can, following the image of a chair that he has in his head, impose something like that form in wood, which was always a chair-in-potential. The potential in the matter, though, in order to be moved to become an actuality, needs to be acted on by a similar actuality in the efficient cause, in this case the idea of a chair in the carpenter's head.[1] With regard to sexual generation, it is the actuality of the male horse that allows it to generate a form by changing the menstrual blood of the female, itself a horse-in-potential, into a horse-in-actuality. (In the terminology of the *Metaphysics*, Aristotle calls the cause and the effect here synonymous, but it amounts to what we have been calling genus-sharing in the biological texts.) To take another example, only what is already actually hot, at least in some sense, can work on a different object, one that is potentially hot, in order to make that object actually hot. The cause and effect again share a genus: the hot. Similarly, musical people can cause, by teaching, other people to become musical. There are some limitations and important qualifications to be made here, but the idea is clear enough: only what is actually (genus) can impose a form on what is potentially (genus) and cause the latter to become actually (genus). Sometimes, to be sure, we need to understand *genus* or *synonymy* fairly loosely, as when motion through the air causes heat in missiles fired from a sling.[2] (This happens in part because, Aristotle says, air is "close to" fire in some sense.) This idea, that actualized forms act on material potentials to cause change, is particularly important for Aristotle as it is the backbone of one of his central arguments against the Platonic theory of forms.

Themistius's objection consists in asking, in essence: if the move from potential to actual requires the action of something sharing a genus with the final actuality, how then do bees come from dead cattle?[3] To change the cattle-corpse, which is bees-in-potential, into bees-in-actuality would seem to require an efficient cause that already had something like the form of bees-in-actuality. But all we had was a dead cow, putrescing with invisible bubbles and heat. As Yoav Meyrav has pointed out, one could simply suppose the emergence of bees to be random and accidental, but that would be to ignore the law-like nature of the event: bees from cattle and wasps from horses, regularly and predictably.[4] The very fact of law-like spontaneous generation, for Themistius, undercuts Aristotle's argument against Platonic forms. Aristotle, Themistius says, "overlooked the many animals that are not born from their likes, in spite of their great numbers. For we see a kind of wasp is born from the bodies of dead horses, bees from dead cows, frogs from putrescence when it becomes sour . . . Unless a suitable formal principle had already been put into nature previously, ready to create any possible species of animal and having found a proper material for creating a certain animal from it, the individual would not have been brought into actuality."[5] How, in short, can Aristotle explain that we get wasps from horses—as we clearly do—if there is nothing wasp-like, in at least some sense, existent in the horse already that could impose the form of "baby wasp" on the material of the rotting body?

Averroes's reply to Themistius involves an invocation of the effects of the sun and the stars, one that he seems to think goes back to Aristotle himself but that is in fact predicated on later developments in the theory of astrology. Framing his discussion of the Themistius passage quoted above, Averroes offers us two related defenses. The first cites what may on the face of it look like a puzzling one-off line from the *Physics*, "man begets man, and the sun does, too," ἄνθρωπος γὰρ ἄνθρωπον γεννᾷ καὶ ἥλιος:[6] "There is no difference between the power which is in putrescent matter . . . and that in the seed, except that [the power] in the seed comes from a being possessing seed and from the sun, whereas that in putrescent matter comes from the sun only. Therefore Aristotle says that man and the sun beget man."[7] The first move he makes here is to liken spontaneous generation to the mechanism for sexual generation, where the form of "man" is imposed on suitable matter by an actual man and the sun. Since the sun is now in play as a cause of formal determination of the species, Averroes is able to use it, in the absence of a parent of the same species, as a cause of formal determination in a spontaneously generated animal. For him the form comes from outside the changing material in both cases, and this, he thinks, answers Themistius nicely. As for how we get regularity of species from certain materials, this is again a combination of material potential and external imposition of form:

> It is the sun and the other stars which are the principle of life for every natural living being, and it is the heat of the sun and the stars which is generated in water and earth which generates the animals [that are] generated from putrefaction . . . As for the heats [that are] generated by the heats of the stars, which produce each distinct species of animals and which are potentially that species of animal, the power present in each one of these heats depends on the amount of the motions of the stars and their reciprocal proximity or remoteness; this power originates from the work of the divine mind, which is like the single form of the single commanding art.[8]

Returning to the common craft analogy, where "the builder's art" is, in a very abstract sense, the actual formal cause of a building for Aristotle,[9] we here have God as the form of all the arts and, one assumes, of all productions of nature. Rather like the Cartesian divinity, which guarantees the laws of nature always and everywhere, the divine mind here acts as an immoveable stopgap for the explanation of the powers of the stars to create species on earth, with or without a parent of the same species.

Now, I said above that Averroes relies on the importation of astrological causes into Aristotle, even if he does not call them that explicitly. In the passage above he only talks about the heat of the stars, not their other astrological qualities, but he does add that species determination is dependent on the stars' "reciprocal proximity or remoteness," which means that their relative positions, at least in some sense, affect the activation of the potentials in earthly matter. Where Averroes pushes this even closer to traditional astrological accounts of stellar causation is in a passage from his *De substantia orbis*, a text whose Latin translation circulated widely in the European Middle Ages. There he offers an explanation that calls upon the most prominent vector of astrological causation, one that we find, for example, in Ptolemy, which is the effect that any particular star's balance of hot, cold, wet, and dry has on earthly matter.[10] In the *De substantia orbis*, Averroes tells us that the stars are able to produce different living creatures, depending on their combinations, precisely because different stars radiate different combinations of the four Aristotelian qualities—a traditional astrological explanation, but one found nowhere in Aristotle:[11] "The heat of the celestial bodies produces generation and bestows vegetative and animal life . . . The statements of those who in former times investigated the activities of the stars have been verified, namely, that some stars impart heat and dryness, some heat and moisture, some cold and moisture, and some cold and dryness."[12] Averroes's explanation does not, to be sure, go into enough detail for us to see whether he has in mind here or in his commentary on the *Metaphysics* any further astrological details beyond position, the four qualities, and speed.[13] Nevertheless, the importation of these new qualities other than heat, as well as the emphasis on planets and stars other than the sun, mark a significant change to the Aristotelian theory.[14] Specifically, they allow for a cause external to the foamy matter as an explanation for species determination, and they therefore circumvent criticisms like the one Averroes perceived in Themistius.

As to why he elaborated the sun and the stars in just this way—effective though they may be against Themistius—it has been suggested that he simply misread Aristotle.[15] We must acknowledge, however, that strange little passage in Aristotle's *Physics* that got Averroes going down this path. Aristotle introduces "man begets man and the sun does too" apropos of the idea that philosophers can abstract out form from matter even if the two are really inseparable in nature. Here he repeats what is a common refrain from the *Metaphysics*, that "man begets man."[16] Uniquely, though, on this and only this mention, he adds the cryptic "and the sun does too."[17] The import of the added line is not immediately clear in its context, but it seems to have to do with an idea he develops most fully elsewhere, in *Meteorology* 1, *De generatione et corruptione* 2.10, and briefly in *Metaphysics* Λ.[18] The issue begins with a puzzle: if fire and air naturally rise, and earth and water naturally tend toward the center of the cosmos, why is the world not just a static ball of earth surrounded by a stagnant coating of water, then of air, then of fire at the outside? His answer is that the constant stirring and heating the sublunar four elements are subjected to by the eternal circular motion of the heavens causes changes that keep things in motion down here and that keep them in motion eternally. Biological generation is likewise dependent on the motion of the sun, partly because generation is dependent on the motion of the four elements and partly because it is dependent on the seasons, which are likewise caused by the sun's motions.

Like the eternality of change in the four elements, the eternal and cyclical motions of the sun also guarantee for Aristotle the eternality of biological generation and are therefore a cause of generation, but in a more abstract sense than an animal's parents are. The heat of the sun is, to be sure, a more direct cause when we think about the seasons for generation, whether sexual or spontaneous, but only in the sense that a certain degree of heat is necessary for menstruation or for the pneuma-rich material of spontaneous generation to undergo its changes from potentiality to actuality. Nowhere does Aristotle take the crucial step that Averroes does, which is to credit the enformation of individual spontaneously generated animals to the balance of hot, cold, wet, and dry radiations from the sun and the stars.

Nevertheless, it appears that Averroes was unaware that he had changed the Aristotelian theory substantially, and indeed, as significant a change as the new theory is, it is a very comfortable fit with Aristotle, and it came couched in a thoroughly Aristotelian framework. This made it an almost unnoticeable tweak to the older theory, and one that seamlessly integrated itself for virtually all of Averroes's Latin readers.

The New Theory Comes to Europe

By the time Aristotle's work on animal generation comes to be widely read in Latin, the Averroan interpretation has taken thorough hold, and Albertus Magnus, the first of our major Latin readers of Aristotle's biology, happily incorporates it wholesale into what he sees as his own faithful reading of the Philosopher.

In fact, the Aristotle that Albertus inherits has been changed at the hands of late ancient and Arabic commentators in many other subtle ways as well, and so, for someone emerging from the forest of Aristotle's biology, coming to read a work like Albertus's monumental *De animalibus* is a heady experience indeed. Not only that, but unlike Averroes's or Galen's commentaries on their predecessors, Albertus declines to follow the time-honored format of plodding through the original piece by piece, quotation then commentary, quotation then commentary. Instead, he interweaves comments, questions, criticisms, and additions into a hybrid text that becomes an impossible-to-untangle mix of Albertus, Aristotle, Averroes, and others.[19] Add to this the clear presence of translation effects (often through multiple translations, such as Aristotle's moving from Greek into Arabic into Latin, possibly with a Syriac or a Hebrew translation somewhere in there as well), and what emerges is, for the student of ancient philosophy, simultaneously both delightful and disorienting.

Not only does Albertus follow Averroes in thinking that spontaneous generation is attributable to the stars, but likewise for Albertus the stars play a role in all generation, and they are also responsible for the birth of what he calls *monstra*:

> non igitur ignoramus, quod sunt quaedam loca in caelo, in quibus cum luminaria convenerint, impediunt etiam in propria et efficaci materia figuram humanam generari: et materia tunc concrescit in horribile monstrum. aliquando etiam e converso concurrunt luminaria et caeteri planetae ad locum, in quo tanta virtus est generationis humanae quod in semine valde difformi contra vim formativam illi semini insitam imprimit formam humanam: propter hoc contingit aliquando porcellos in capite praeferre figuram hominis, et foetus vaccarum similiter . . . haec igitur est causa et non alia, quod etiam in lapidibus vaporabiliter in materia coagulatis imprimitur figura hominis vel alia de figuris specierum quas producit natura, aut pingendo tantum, aut etiam figurando totum elevando in toto vel in parte, maxime cum sit huius efficiens in onychinis propter materiae maiorem mollitiem, ut diximus supra.[20]
>
> We are not unaware that there are some places in the heavens where, when the stars come together, they prevent the generation of the human form, even in material proper and suitable, and then the material grows into a terrible monster. Also conversely, at other times the stars and other planets come together in a place in which there is such a power for human generation that it imprints a human form in seed of a very different form, against the inherent formative virtue of that seed. Because of this it sometimes happens that pigs have a human face on their heads, and the offspring of cows likewise . . . This is therefore the cause (and there is no other) that likewise in stones there is impressed the figure of a man or of another species which nature produces, hardened in the material by vapour, sometimes painted in, sometimes carved in, raised up either in whole or in part. This happens especially in onyx because of the great softness of its material, as we said above.

Notice that Albertus cites the positions of the stars in the sky as his primary cause of these effects.

In addition to *monstra*, the stars have such generative power sometimes—and that power is so keyed to particular human and animal forms—that these forms are engendered not only in the living flesh of different species of animals but even in the nonliving matter of stones. He then goes on to tell us about one such stone he has seen in Cologne, with the figures of the heads of two young men on it. He describes it in exquisite detail, so minutely that we are today actually able to identify the exact—and still-extant—stone that he had in mind. It is, in the event, an exquisitely carved (and, alas for Albertus, manmade) Ptolemaic cameo now residing in the Kunsthistorisches Museum in Vienna. That Albertus could see the cameo as a product of nature is more than a little surprising, and it serves perhaps as a timely reminder of just how very different the world he lived in was from our own.

Historians of science and religion will be familiar with the "divine watchmaker" argument for the existence of God, usually attributed to William Paley (1743–1805),[21] who argued that if you found a simple stone on a heath, you might suppose that it had always been there, but that if you found something as complicated as a working pocket watch—even if you didn't know what it was or what it was meant to do—you should suppose that it was made for some purpose by an intelligent being. The whole wonderful, complicated, and flawlessly ticking universe, on this argument, must likewise have a maker and a purpose. In some ways, we can see Albertus's natural mechanism for the generation of cameos as a fascinating inverse of the divine watchmaker: the generative power of the stars is so great that things that we now know to have been purpose-made by people, get taken by Albertus for natural stones found on heaths. Here the conceptual context in which the cameo was situated for Albertus goes a considerable distance toward explaining how he could see the thing as anything other than a human creation. For him, the mechanism he is proposing for the production of the cameo covers not just this one class of object, remarkably beautiful, detailed, and life-like as these are, but also some kinds of fossil—equally beautiful, detailed, and life-like. Although Albertus thought that many (most?) fossils were formed when the bodies of living animals succumbed to a *virtus mineralis* in the water that converted them

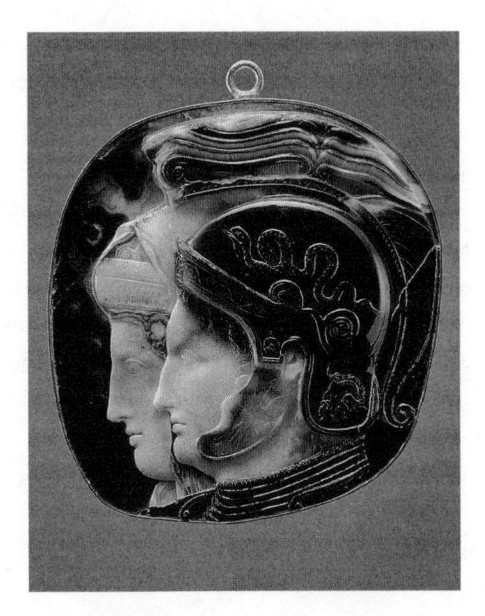

Albertus's stone found on a heath.
Image courtesy of the KHM Museumsverband, Vienna

to stone, he floats a different theory early on in his book on minerals that explains how the fossils called "moonshells" are produced, which is as animals generated spontaneously within already existing stones:

> animalia quae testudines vocantur, frequentissime cum suis conchis generantur in lapidibus: et hoc frequentissimum est in lapidibus qui inveniuntur Parisiis,[22] quibus plurima foramina sunt habentia figuras concharum testudinum, quas quidam lunares appellant. causa enim huius est humiditas quae ad locum illum evaporavit, et a circumstante materia, in se convoluta est, et exterius primum constans intra se circumacta, spiritum vitalem accepit.[23]

> The animals which are called testacea are very frequently generated with their shells in stones, and this is very common with stones found at Paris, in which there are many holes with the figures of the shells of testacea, which some people call "moonshells." The cause of this is the humidity that has evaporated toward that place and that has been rolled

up into itself from the surrounding matter, and, establishing first the outer [part] by being turned within itself, has received a vital spirit.

That fossils were sometimes or always formed as animals spontaneously generated in stone, while not the commonest explanation, is far from unheard of in antiquity and the Middle Ages. We have a single example in Galen's *On Antecedent Causes*, a text that only survives for us in a fourteenth-century Latin translation (and is, for the record, the passage that started me on this entire project). At one point in an otherwise very interesting argument about causation, Galen rather pedantically discounts one type of incidental "cause" as being especially meaningless. This is the air though which a person moves when making something (in his example, a carpenter turning a bowl). It is true that the bowl could not be made if the carpenter had no air, no intervening space, through which to move in making the bowl, but it is hard to see this intervening space as a "cause" of the bowl in any meaningful way. In the midst of his discussion, Galen has us imagine what it would be like if the carpenter could somehow bring the bowl into existence, not by working on it externally, but by permeating the entirety of the wood, *in medio lapidum, aliquando autem in terra generantur animalia eundem locum occupante conditore cum eo quo fit*, "this is, after all, how animals are generated in the middle of stones or sometimes in earth, when their maker occupies the same space as that which it makes."[24]

So no stellar influence for Galen, but spontaneous generation (and, one assumes, a rather unpleasant death) within the very stone itself. For Albertus, the generation of animals in rock, while not the cause of all fossils, is certainly possible under some circumstances, and as we've seen, he has a host of animal and human images being created by the same force, but without with the addition of a *spiritus vitalis*, and so entirely unliving. The range of possibilities for Albertus is wide: living animals in stones, stones that look like animals and people down to the finest detail but which are not and never were living, and even images of animals that are produced in and on other animals. Albertus tells us about some images of serpents *optime factorum ita quod nec figura defuit oculorum, cum tamen essent valde parvi*, "so perfectly made that not even the image of the eyes was missing even though they were very small."[25] These appeared on an oyster shell, caught in the belly of a fish, that was given to him as a gift by the Prince of

Castile (quite possibly Alfonso X, of the Alphonsine tables). In this passage we see the same remarkable attention to detail that Albertus brought to his description of the Ptolemaic cameo, and here we get the added bonus of an explicit claim to firsthand experience of the thing. It should be said that no one doubts that Albertus actually saw the cameo (it was housed at the time in the very city in which he lived and wrote for much of his life), but in the case of the oyster shell, he goes so far as to tell us its entire provenance and even to explain why he no longer has it in his possession (should we be inclined to ask). He takes his reader from the moment he acquired the thing, "when I was in Paris," to *hanc autem concham ego multo tempore habui, et multis ostendi, et postea eam misi pro munere in Teutoniam cuidam*, "I had this shell for a long time and showed it to many people. Later I sent it as a gift to someone in Teutonia."[26] Multiple witnesses, many showings, its value as a gift—all attesting to just how remarkable the shell was.

Finally, there is one practical upshot of this heavenly generation of animal forms in stone, which is that the stones themselves, impressed by the astral influence, continue to hold that astral influence within themselves and can be used medicinally and to work wonders. Related to this is the human practice of making amulets, which is only successful, Albertus tells us, if the timing of the manufacture is able to take advantage of the correct stellar influences,[27] signaling again the thoroughgoing incorporation of an astrological influence into the Aristotelian edifice.

Questioning the Evidence

Albertus often follows Aristotle in reporting on the spontaneous generation of eels, but he also questions him on key points. We saw earlier a number of places where Aristotle's own personal observations were invoked by him as proof or example of spontaneous generation. One kind of such claim takes a slightly less personal, but still very effective, form. In the *History of Animals*, book 6, Aristotle points out, as evidence that they are spontaneously generated, that

> ἐν δὲ τοῖς ἐντόμοις καὶ τοῖς ἰχθύσιν ἐστὶ τὰ μὲν ὅλως οὐκ ἔχοντα ταύτην τὴν διαφορὰν ἐπ' οὐδέτερον, οἷον ἔγχελυς οὔτ' ἄρρεν ἐστὶν οὔτε θῆλυ, οὐδὲ

γεννᾷ ἐξ αὑτοῦ οὐδέν, ἀλλ' οἱ μὲν φάσκοντες ὅτι τριχώδη καὶ ἑλμινθώδη πρασώδη τ' ἔχουσαί ποτέ τινες φαίνονται, οὐ προσθεωρήσαντες τὸ ποῦ ἔχουσιν ἀσκέπτως λέγουσιν . . . ᾠὸν δ' οὐδεμία πώποτε ὦπται ἔχουσα.[28]

among insects and fishes there are some that actually do not have [a] distinction according to the two [sexes], thus the eel is neither male nor female nor does it generate anything from itself. Those saying that they have sometimes been seen to be pregnant[29] with hair-like or weedy worm-like things speak ignorantly without carefully observing where these things are [in the eel's body] . . . And no [eel] has ever yet been observed having an egg.

Eels were commonly thought to begin as worms in the mud (specifically, as the worms called "earth's guts"; in Greek, γῆς ἔντερα), so finding an eel pregnant with anything even vaguely vermiform would be a compelling argument against the common belief. But as Aristotle says, the people who thought they saw such things were mistaken about where in the eel those worm-like things were, hinting, I assume, that they were probably just parasites (which may themselves have been spontaneously generated in the body of the eel).

Albertus, however, has his doubts:

in isto pisce dicitur non esse mas et femina, quamquam iam bis a fide dignis audierim quod duae anguillae captae sunt in Germaniae partibus quarum utraque multas habuit filaris quantitatis in utero, et matribus occisis ex ventribus earum multae egrediebantur.[30]

In [the eel] there is said not to be male and female, although I have heard twice from men worthy of trust that two eels have been captured in parts of Germany each of which had many filaments in its uterus, and the wombs being opened, from their bellies many came out.

There are several features here worthy of attention. In telling us that his sources are worthy of trust, Albertus is signaling his own faith in their reports and, of course, exhorting his reader to follow his lead. Secondly, he locates his sources notably close to home (it is a quirk of Albertus's writing that he often adds "there are many of these in Germany" when he mentions eels),[31] again, one can only assume, with an eye to plausibility. Finally, Albertus goes to no small lengths to emphasize the location of the filaments. He tells us that they were found in the uterus, and that they emerged from the belly on dissection of the wombs, just to be sure. He is quite clearly here pointing his passage directly back at Aristotle's statement that people who make such claims speak ignorantly, without careful attention to where in the body the filaments are found. One gets the distinct impression that Albertus pushed his sources, speaking with them and asking pointed questions, on just these details and specifically in light of Aristotle's assertion about the location of the filaments in the body. He is entirely confident that Aristotle made a mistake in this matter, and he wants his reader to share in that confidence.

But there is another curious aspect about Albertus's assertion here, one that shows up the full complexity of the project on which Albertus sees himself as working. This claim comes only in book 24 of *On Animals*, after he has mentioned eels many times before, and having repeated the claim several times that they have no sexes:

> invenitur tamen animal de quo non adhuc apparuit mas aut femina, et hoc est anguilla.[32]

> An animal is found, however, for which no male or female is as yet apparent, and this is the eel.

> omnes autem masculi piscium semen habere inventi sunt, praeter quaedam genera, quae hahaloz dicuntur; de quibus supra fecimus mentionem, in quibus nec ova nec sperma invenitur: et de istis generibus est anguilla.[33]

> All males among the fishes are found to have semen except some genera which are called hahaloz—we have made mention of these above—in which neither egg nor sperm is found; the eel is of this genus.

Kitchell and Resnick point out, plausibly, that Albertus is probably just mangling the Greek word for eel, ἔγχελυς, with his *hahaloz*, after all his spelling (or his source's) for the Greek *enchelys* is pretty flexible: he gives it in different places as *enchelyz*, *encheliz*, *enchelos*, and *encheluiz*, so why not *hahaloz*?[34] Still, the important thing is that in these two passages we see what appears to be quite the opposite claim to the later one in book 24, that two reliable fisherman have reported what Albertus has no doubt are females.

A similar, but even more compact string of conflicting claims happens on the topic of whether eels have eggs. Over the course of just a few pages in book 6, Albertus claims, on the one hand, that *encheliz, quod est anguilla, tempore pluviae in gutture ova invenitur habere aliquando*, "the encheliz, that is the eel, in rainy weather is sometimes found to have eggs in its neck,"[35] which would seem a rather shocking claim given his earlier discussion of the wombs and uteri of eels (or it at least should be shocking if we make the natural assumption that he thinks these are eel eggs). It takes him a few pages to offer us the correction: *ova etiam, quae inveniuntur in eis tempore longae pluviae, non inveniuntur in loco ovorum in ventre vel in matrice, sed in collo; et ideo sunt superfluitates granulosae similes ovis, sed non sunt ova secundum veritatem*, "the eggs which are found in them during periods of long rain are not found in the place for eggs, in the belly or the womb, but in the neck, and they are granular growths like eggs, but they are not eggs in the true sense."[36] So he does offer us a correction (this time not after a full twenty-three books but instead in the space of just a few pages), but the effect on the reader is striking all the same. No one in the thirteenth century who knows anything at all about eels is going to pass over that first sentence without raising an eyebrow, and yet Albertus lets it slide, correcting it at leisure, and only when it occurs to him to do so.

Perhaps this apparent equivocation is just a characteristic of Albertus's writing style, which at the best of times strikes one as hasty (his handwriting was also legendarily atrocious).[37] Perhaps it is an artefact of his sometimes copying passages from Aristotle or Averroes for one purpose, commenting on one aspect or another, but failing to scrutinize the whole every time—if he eventually gets around to putting down the correction or the concern, well, his reader has it all there somewhere, so what, perhaps, is the harm?

Recall now a passage that we saw in an earlier chapter, where Aristotle discusses the possibility that some species may have females but no males and so may generate parthenogenically: εἰ δ' ἐστί τι γένος ὃ θῆλυ μέν ἐστιν, ἄρρεν δὲ μὴ ἔχει κεχωρισμένον, ἐνδέχεται τοῦτο ζῷον ἐξ αὑτοῦ γεννᾶν. ὅπερ ἀξιοπίστως μὲν οὐ συνῶπται μέχρι γε τοῦ νῦν, "if there is a species which is [entirely] female and has no distinct male, it may be possible for this animal to generate from itself. This has not been reliably observed thus far, at any rate."[38] As we saw, after floating this possibility Aristotle goes on to name a few kinds of fish that are good candidates, and he ultimately leaves the question open. But if we look to the section in Albertus's *On Animals* that paraphrases this passage, we see considerably more confidence on the part of Albertus:

> si autem esset aliquod genus animalium femininum et non esset masculus distinctus a femina, sed haberet in eodem individuo utrumque sexum, forte posset generare talis femina per seipsam: sed talem speciem animalis numquam vidit aliquis hominum veridicorum, de cuius sermonibus sit multum curandum.[39]

> If, however, there is some genus of female animals and it does not have a male apart from the female but has each sex in the same individual, perhaps it is possible for such a female to generate by herself. But no one truthful and whose testimony should be given much credence has ever seen such a species of animal.

We got the distinct impression from Aristotle that his conclusions were preliminary and he was open to the possibility that some animals were female-only species, whereas Albertus dismisses the possibility outright, with a double slander about both the veracity and the care of whomever he has in mind as claiming it to be true.

Part of the problem here is the same issue that we mentioned in connection with aphids previously: one cannot, logically speaking, prove the absolute nonexistence of a thing. One can only rule it out so often that one eventually decides the thing in question is nowhere to be found (indeed, this is just what we've finally come to do with spontaneous generation as a whole). But as we've seen, there are, historically, any number of positive claims to the experience of spontaneous generation, and it is these observations that do all the real work of convincing Aristotle, for example, that eels are spontaneously generated. When Aristotle says that earthworms have been dissected and eels found within them, he is not being disingenuous, and we have seen him carefully qualify the amount of confidence he has in a wide range of his claims. What he does not do, however, is try to actually generate eels or other animals from scratch.

It turns out, though, that we have a later example of someone who does.

Giambattista Della Porta's *Magiae naturalis* of 1589 is a veritable *tour du monde* of wondrous claims. To read it is to be transported vividly back into a world where there really is a great deal more in heaven and earth than we now dream of. Having said that, Della Porta is no simple courier of fantastic tales. In fact, he frequently puts the claims of his forebears (of whom he has, it should be noted, a cultivated and encyclopedic knowledge) to the test—and I note that Della Porta was the first person to claim to have empirically disproved the old story that garlic would ruin magnets.[40] No less so with spontaneous generation:

> in anguillis sexus foemininus, masculinusque non est, nec per coitum generantur, nec pariunt ova, nec vero capta unquam est, quae aut semen genitale, aut ova haberet, nec meatus vulvae accommodatos scissura ulla ostendit.[41]
>
> Among eels there is no feminine or masculine sex, and they do not generate by mating, nor do they lay eggs, nor even when captured do they have semen or eggs nor does any dissection show suitable passages to the womb.

So far, so much Aristotle. But he then continues: *ex putridis etiam rebus generantur. compertum est enim mortuo equo, et stagnis iniecto paulo post innumerabiles anguillas visas esse*, "they are generated from rotting things. For it is a fact that when a dead horse is thrown in stagnant water, a little while after innumerable eels will have appeared." The phrase that I am translating here as "it is a fact that," *compertum est*, could also be rendered as "it has been ascertained," "proved," or even "witnessed." Whether this is a claim to personal or proximal experience is not entirely clear, although Della Porta's full confidence in its veracity certainly is. And as he goes on to develop his account of the eel, he makes an observation claim that is even more direct. After telling us of the role of seaweed in the generation of eels, he adds: *ob id nos amicum novimus, qui ligneis vasis aqua plenis, algis, et quibusdam fluminum herbis immissis, et lapidis pondere oppressis subdio, paucis diebus anguillas generabat*, "on that point, we know a friend who, using wooden vessels filled with water and seaweeds with some other river plants thrown in, and with stones pressing heavily down under the open sky, after a few days generated eels." Note the emphasis on the careful closing off the vessels with heavy stones—a precaution against experimental contamination?—and again the emphasis on close-to-hand, experiential knowledge.

Della Porta's entire section on spontaneous generation is littered with experience claims: *quotidiana nos videmus experientia multa*, "every day we see many experiments," *ipse autem pluries expertus sum*, "I myself have often had the experience," *idem evenire videmus . . . atque ab amicis idem factitatum accepimus*, "we have seen [horse hair turn into serpents] . . . and we have [heard] the same frequent experience from friends,"[42] and a host of others. But before we go thinking that Della Porta doth protest too much, we should keep in mind just how diligent and forthcoming he could be about his failed experiments. For example, he seems to have taken an especially keen interest in producing mushrooms, telling us:

> quid referam quot fungorum et tuberum species produximus, ut quot putrescentium rerum misturae essent, tot variae species emanarent? quae omnia iam huc afferrem, si ad normam redigere potuissem, vel ea enata essent, quae proposueram, sed nequicquam quaesita, oriebantur.[43]

> What shall I report of how many species of mushroom and truffle we have produced? That for however many mixtures of putrefying material as there had been, so many different species came forth? I would have reported all of these that I have produced had I been able to put them in some order or if they had been fruitful but my inquiries have amounted to nothing.

How many species, how many mixtures? He comes across as someone who is frustrated in the face of not having been able to find an order to impose on his all-too-wide-ranging results. He adds that *Deo dante*, God willing, he will eventually write something on the subject, but that time and circumstance do not currently permit.

He is quite open about not always succeeding in his tests, and he is certainly not credulous of everything he reports. On the production of asparagus from wild ram's horn, he reports the opinions of Didymus and Pliny, but then says he has had no luck in trying the recipe himself: *etsi a nobis facto frequenter periculo, non evenerit, ab amicis autem audivimus ex tenero illo sato intra arietis cornu, contento, experientiae successisse,* "and this experiment being tried frequently by us, it has not turned out, but we have heard from friends that from that soft (part) begotten enclosed inside a ram's horn, they succeeded in the trial."[44] These same friends, and Aristotle, report that ivy comes from stag horn, *etsi mihi periculum facere non contigerit,* "but I have not had the opportunity to do the experiment."[45]

The Birth of an Underclass

I have remarked several times on the redefinition of perfect versus imperfect animals in the medieval period. Whereas for Aristotle animals and eggs that changed in form or size after separating from their mothers (insects, fish eggs) were imperfect (*atelés*),[46] in Albertus and then in Aquinas after him, perfect and imperfect take on whole new meanings, and the distinction settled on by Aquinas would become a significant one for the future of spontaneous generation.[47] For Aristotle, animals generated spontaneously were classed only as animals that generated spontaneously; they did not share supervening qualities beyond this. By redefining the terms and eventually classing all spontaneously generated animals together under the banner "imperfect," Albertus, Aquinas, and the later tradition would start to see fundamental differences between sexually generated animals (which were larger, more complex) and spontaneously generated animals (small, simple), and these differences would determine possibilities for later thinkers on what animals may or may not be candidates for spontaneous generation.

As we shall see, it is Aquinas who finally establishes the terminology for the later tradition, but Albertus certainly moves us in that direction. For Albertus, perfect and imperfect signal degrees of complexity, and ultimately degrees of likeness to humans, since, as he tells us in book 21, humans are the "most perfect" animal. This perfection establishes their position atop the great hierarchy of animals, which they occupy on the grounds of their faculties and "virtues," *virtutes*. These include the obvious—we have the faculty of reason, where nonhuman animals like monkeys and (as he supposes) "pygmies" only have a semblance of reason. But he also appeals to a host of other human features, including our truly multipurpose (he might almost say omnipurpose) hand, an organ possessed by no other animal.

One surprising source of human perfection is the symmetry of our physiology, and he gives us a description that—a little surprisingly to my mind—vividly recalls Da Vinci's Vitruvian man. Albertus appeals again and again to his theory that the virtues of the soul in any animal, its strengths and weaknesses alike, can be discerned from its physiology; therefore fish, which are bilaterally symmetrical, are more "ordered in their perceptions" than are octopus, which have arms growing and waving about, any which way.

As Albertus goes through his list of animals and their degrees of perfection, he ranks genera and species from most perfect to least. Beginning with man, pygmies, and monkeys, he works his way down through quadrupeds, birds, and aquatic animals, to "creeping animals" and insects, ending finally with the worm and the sponge. The criteria for perfection that he cites again and again include the powers of the soul for memory and recollection, which he ties to a species' ability to be taught new things and which he underpins with a nicely elaborated theory about the psychology of learning and remembering. Prudence and shrewdness (*prudentia et sagacitas*) are significant, often playing out as an animal's ability to hunt or avoid predators. Relevant, too, are the number and quality of senses possessed by each genus. His final animal, the sponge, which he thinks of as nearly being a plant, is said to be "of the greatest imperfection," *imperfectionis igitur maximae sunt et extremam naturae animalium habent participationem*, "and they have the least participation in the nature of animals."[48]

In some places Albertus talks about the perfection and imperfection of organs, independently of the perfection of the animal to which they belong. In animals with lobed lungs, the composition of these organs is said to be *completa et perfecta*.[49] Animals with blood and with hearts in the middle of their bodies are said to be "more perfect," *perfectiora*.[50] The sea squirt is called *valde imperfectum*, "very imperfect."[51] There are passages where he seems to be calling sea squirts, sponges, and other simple marine animals imperfect, and he says of testacea *haec igitur sunt imperfecta valde*, "these [animals] are truly imperfect,"[52] and elsewhere he calls these and insects *imperfectiora*, "very" or "more imperfect."[53] In one instance he refers vaguely to some animals from the sea as imperfect.[54] In some passages, as at 15.41, he seems to equate perfection with size, although he is not entirely consistent on this.

When it comes to discussing sex differentiation,[55] Albertus distinguishes between those animals in which the sexes are distinguished perfectly (all blooded animals) and those in which they are distinguished imperfectly. He says that perfect sex distinction is found only in blooded animals, and he seems to suggest that nonblooded animals (he clearly has insects in mind) are imperfectly distinguished with regard to sex and that this applies to them as a class, but he does admit that in some of them there are distinguishable sexes, and they generate by copulation, though their offspring is not (at first) of their same genus. These and other passages make it clear that if an animal is blooded and therefore necessarily also has a clear sex distinction, that animal is perfect.[56] But Albertus's deployment of imperfection is not so simple.

A notable feature of Albertus's uses of the category of imperfection is that imperfection is frequently situated only with respect to some particular feature of an animal. Sometimes he applies the term to an animal's means of locomotion, or to the number of senses it has, or, as we just saw, to how clearly distinguished its sexes are. When he called testacea "truly imperfect," above, he was concluding a discussion of their means of locomotion (or lack thereof), and his uses often seem to fit this pattern of being aimed at one feature or another. At the same time, there are a great number of instances in which he simply waves his hands, referring vaguely to imperfect animals but not really telling us what he has in mind. On those few occasions when he gets more explicit, things look promising:

> alia autem cum imperfecta sint, et praecipue ea quae imperfectiora sunt inter ea, imperfectius etiam habent suae generationis principium: propter quod etiam generantia talia aut nullo modo coeunt aut non spermatizant in coitu.[57]

> The others, though, since they are imperfect, and especially those among them that are more imperfect, have a more imperfect principle of their generation. Because of this, then, such ones that do generate do not mate in any way or else do not eject sperm when mating.

And at 16.75, he says: *a defectu principiorum generantium fiunt animalia imperfecta*, "imperfect animals are made by a defect of the principles of generation," but he then goes on to discuss egg-laying animals in this category, so he cannot mean here what Aquinas will later mean, limiting imperfection to spontaneously generating creatures. Again, imperfection is often for Albertus relative to a particular trait, and there are apparently degrees of imperfection that extend even to animals we should expect to be called perfect, such as birds, which are both sanguineous and perfectly sex-differentiated.

Our confusion is not helped when we come to the chapter titled "On Imperfect Animals, and the Reason for and Degree of their Imperfections," for there Albertus limits himself to only two animals, the earthworm and the sponge, and he explains that they are imperfect not because they are spontaneously generated (he does seem to think the earthworm is so generated, but he is silent on the sponge) but because they have such limited sensation, the earthworm having only touch, the sponge nothing at all. That the sponge is immobile makes it even more imperfect than the earthworm, but even the earthworm is imperfect insofar as it moves by contraction rather than walking.

Nevertheless, the old Aristotelian sense of a perfect animal, one that is born resembling its parents, does also make the occasional appearance, as at 21.47, where Albertus describes what is likely a butterfly or moth, saying that when it emerges after pupation, *eo quod tunc tandem ad suae speciei perfectionem est reversa species haec animalis*, "then at last this animal species has returned to the perfection of its species."

Thus, although we can see Albertus moving toward a definition of *imperfection* that conjoins simplicity, size, and spontaneous generation, he remains remarkably generous in his usage of the term. Where we find the clearest version of the definition that finally settles down for posterity, is in Aquinas:

> animalia enim perfecta videntur non posse generari nisi ex semine; animalia vero imperfecta quae sunt vicina plantis, videntur posse generari et ex semine et sine semine . . . et hoc rationabiliter accidit. quia quanto aliquid perfectius est, tanto plura ad eius completionem requiruntur. et propter hoc ad plantas et ad animalia imperfecta, sufficit ad agendum sola virtus caelestis. in animalibus vero perfectis requiritur cum virtute caelesti etiam virtus seminis. unde dicitur in secundo physicorum quod homo generat hominem et sol.[58]

> For perfect animals it is clear that they cannot be generated except from seed, but for imperfect animals which are close [in kind] to plants, it is clear that they can generate with seed or without seed . . . And this is rational, for however more perfect a thing is, that much more is required for its completion, and because of this the power of the heavens alone is sufficient for acting in plants and imperfect animals. But in perfect animals a seminal power is required with the power of the heavens. Thus it is said in *Physics* 2 that man generates man and the sun does too.

Thus it becomes definitional that an imperfect animal is one that can be generated by the sun alone, whereas a perfect animal requires a parent and seed. In addition to sharing a mode of generation, spontaneously generated animals now all share a position at the bottom rung of the great chain of animal species. Everything that is sexually generated is above them.

Final Causes and Chance

If spontaneously generated animals are lower down the ladder than sexually generated animals, that does not necessarily mean that they do not still show some order and purpose. We have already seen how the sun and stars had come to be used to explain the generation of specific forms in matter. Closely related to this was the question of the final cause of such creatures. For an acorn, its final cause, the thing it strives toward in every aspect of its development once germinated and rooted, is to become an adult of its species. Although this is closely related to the question of what the form of an acorn is, it is conceptually a distinct question, and one that needed to be worked out for spontaneously generated animals as well. After all, even Aristotle admitted that forms could sometimes come to be by chance, which is to say, without or in spite of a proper final cause.[59] These things have a form, and may have started with the imposition of a form, but due to accidents and chance, they did not develop in the way they should have. The question, then, of how to reconcile the spontaneity of spontaneous generation with the purposive development of an individual of a particular species needs clarification.

Aquinas argues that spontaneously generated animals do have a final cause, even if in some respects their generation may be due to chance (why this individual? why now? why this bubble and not the one next to it? why . . . , etc.?). Just as with the imposition of living form on suitable matter, it is the stars that solve the problem:

> sed sciendum est quod nihil prohibet aliquam generationem esse per se, cum refertur ad unam causam, quae tamen est per accidens et casualis, cum refertur in aliam causam. sicut in ipso exemplo Philosophi patet. cum enim sanitas ex confricatione sequitur praeter intentionem confricantis, ipsa quidem sanatio, si referatur ad naturam, quae est corporis regitiva, non est per accidens, sed per se intenta. si vero referatur ad intellectum confricantis, erit per accidens et casualis. similiter etiam generatio animalis ex putrefactione generati, si referatur ad causas particulares, hic inferius agentes, invenitur esse per accidens et casualis. non enim calor, qui causat putredinem, intendit naturali appetitu generationem huius vel illius animalis, quae ex putrefactione sequitur, sicut virtus, quae est in semine, intendit productionem talis speciei. sed si referatur ad virtutem caelestem, quae est universalis regitiva virtus generationum et corruptionum in istis inferioribus, non est per accidens, sed per se intenta; quia de eius intentione est ut educantur in actu omnes formae quae sunt in potentia materiae.[60]

> We know that nothing prevents any [kind of] generation from being a proper process with respect to one cause and also being by accident and random with respect to a different cause. This is clear in that example from Aristotle: when health follows a massage and this happens independently of the intention of the person doing the massaging, this process of healing, when considered with respect to the nature that governs the body, happens not by accident but is properly directed. But if it is considered with respect to the intent of the person giving the massage, [the result] would be by accident and random. Similarly, the generation of animals that are produced by putrefaction, if considered with respect to specific causes acting here below, it will be found to be by accident and random. For the heat that causes putrefaction does not aim, by a natural desire, at the generation of this or that animal which comes from putrefaction in the same way as the power that inheres in the seed aims to produce just this species. But if we consider it with respect to the celestial power that is the universal ruling force in these lower

> regions, [the generation of a particular animal] is not by accident but is properly intended because, concerning its goal, this is that all forms which exist potentially in matter be brought forth in actuality.

This last sentence is a significant addition, but it would enable scholastics, in the face of Platonic criticisms such as those of John Buridan,[61] to argue that external forms are not required for spontaneously generated animals to have a proper final cause. The world is simply made such that the potential for life—any kind of life—wants to be actualized in any matter that harbors that potential. We see this same idea—although with an even more grandiose spin applied to it—reappear three and a half centuries later, as the central timber of Fortunio Liceti's massive, influential, and exhaustive *On the Spontaneous Generation of Living Things* (1618).[62]

Liceti follows some of our earlier authors in focusing on the fact that one always gets bees from calves and wasps from horses as proof that spontaneous generation is not indeterminate or random, and that something to do with the potential bees and wasps inheres in the matter or the cow and horse. As part of establishing his argument, Liceti situates spontaneous generation in a surprisingly complex and exhaustive taxonomy of generation:

> octo igitur ad summum sunt animantium generationes, per semen corporeum, per virtutem seminis incorpoream, per sexuum indistinctionem, per divisionem, per favationem, per semen simul et citra seminis interventum a duplici agente proximo, per creationem, atque spontanea generatio nuncupata: quas omnes, una excepta creatione, variis in locis distinctas agnoscit Aristoteles.[63]

> And so there are in total eight kinds of generation for living things: [that] through corporeal semen; through the incorporeal virtue of semen; through an indistinct sex; through division; through honeycombing; through semen but at the same time also by a second proximal agent without the intervention of semen; through [divine] creation; and through what is called spontaneous generation. All of them, with the one exception of creation, are acknowledged in different places by Aristotle.

Even the careful reader of Aristotle could be forgiven for not recognizing that Aristotle had enumerated seven of these different kinds of generation, and so to clarify the terminology here, let us briefly explain Liceti's list of the different means of generation. They are (a) corporeal semen: this is how most animals and plants and all humans are produced, the semen of the male enforming the matter of the female; (b) incorporeal seminal virtue: this is how Liceti understands the mode of generation in those insects in which the female inserts something into the male and "mating takes a long time" (Aristotle had puzzled over this situation but did not conclude that there was an "incorporeal virtue of the semen" at work);[64] (c) an indistinct sex: this covers Aristotle's *erythrinus* and *channa*, for example (Liceti does not mention bees in this connection, but my guess is that they also go here); (d) division: includes plants and insects that can be divided in two, and both halves will live;[65] (e) honeycombing: this is how mussels and some other testacea are generated; (f) by semen and at the same time without a seminal intervention: this is solely for the generation of humans (Liceti says that only God can bring the rational soul into existence; Aristotle had said only that the rational soul had to come from "outside");[66] (g) by creation: this is what God did when he created the world and all life; and finally, (h) spontaneous generation: although notice that Liceti is now distinguishing this from honeycombing ([e], above), whereas Aristotle had seen honeycombing as just one of the ways spontaneous generation happened.

As I presented him in the first chapter, Aristotle divided generation into sexual and spontaneous as his main two categories, with a few caveats concerning the puzzles with bees as well as the uncertainties surrounding the *erythrinus* and *channa*. True, there are different means of sexual generation (though he never really enumerates them in a list as Liceti does): insects generally differ from other animals, and bees may contain both sexes in one individual. So, too, there are many different means of spontaneous generation, some animals coming from foam but mussels honeycombing (he agrees with Liceti that they emit a quasi-seminal fluid, but Aristotle doesn't actually say that this significantly distinguishes their mode of generation from what is properly "spontaneous").

What Liceti is doing in this passage is drawing a much starker line between Aristotle's subclasses, if only for a moment (and perhaps to indulge in a bit of pretty scholasticism).[67] What stands out in this list, though, is that for the Christian Liceti, God clearly had an active role in the generation of life only at the beginning, when he created the world, with the one exception that he continues to have an active and indispensable role in the generation of human rational souls. The souls responsible for human nutritive and perceptual/motor functions (growth and nutrition, on the one hand, and sense, emotion, and mobility, on the other) are received directly from the semen of the male parent without divine action, just as happens in the other animals.

For Liceti, the form and the soul of the spontaneously generated animal preexist in matter. This is a little more specific and perhaps a firmer claim than Aristotle had made. As he says in a chapter summary: *formam, et animam, a qua ut ab efficiente univoco proxime generantur animantia sponte nascentia, praeexistere in eorum materia, seu in cadavere, ut in vase, nihil ad eam conferendo, colligitur ex rei natura*, "the form and the soul, from which spontaneously generated organisms are generated as by a proximate, univocal, efficient [cause], preexists in their material or in the dead body as in a container. Nothing needs adding to it; it is brought together by the nature of the thing."[68] He elsewhere specifies that the material from which spontaneously generated beings come into existence must in all cases have been previously living, which marks a break from Aristotle. The not-yet-living soul of the spontaneously generated organism is such that *porro inest huiusmodi anima in materia illa, ut in vase, non aliunde, quam a pristino vivente*, "such a soul is in that material at a distance, as in a container—not any [material], but that which was previously living."[69]

His constant repetition of the phrase "as in a container," *ut in vase*, marks another distinction from Aristotle,[70] flagging a significant position with respect to which we will see later authors falling on one side or the other. It marks a very nice fit, as I hinted at earlier, of Aristotle with something very like Augustine's seminal principles.

Liceti now goes on to specify that spontaneous generation in plants is the vivification of a nutritive soul only, and in animals of a desiderative soul (necessarily bringing with it the nutritive capacity). He also nicely distinguishes the mechanism of spontaneous generation of living things from the generation of any kinds of nonliving things (metals and rocks in the earth).[71] Finally, he summarizes the four causes of spontaneously generated beings. The material cause is simply the material in which the new life emerges (it uses that material as food and as physical body). The immediate efficient cause is

> anima illa nonvivificans praeexistens in materia, ut in vase, in ea vel primum genita, vel relicta a pristino vivente, sub inferiori gradu; aut quia prior anima in eam alterius speciei degeneraverit; aut quia quae prius erat forma vivificans subditam sibi materiam, in ea maneat ociosa, torpensque lateat ex necessitate materiae.[72]

> that nonliving soul [that is] preexistent in the material, as in a container, either born for the first time in [the material] or a leftover from the previously living [material] according to an inferior type, either because the earlier soul of another species in [the material] has degenerated, or because what was previously the form that vivified the underlying material must remain idle in it and lie sluggish because of a material necessity.

Liceti explains how the form of the new organism is due to the material in which it was created when that material was stimulated by the heat of the atmosphere.

The final cause is "in one sense, the existence itself of a living thing," *est tum ipsum esse viventis*, "in another sense the perpetuation in species of the generating soul according to its nature," *tum animae generantis perpetuatio in specie pro modo naturae suae*.[73] But what does he mean by the perpetuation in species of the generating soul according to its nature? The idea looks a little like what Aquinas had in mind, that the purpose of spontaneous generation was to realize in actuality all the potentials for life that inhere in matter. But Liceti goes further than this, claiming that the continual generation of every individual species, the constant cyclical coming-to-be of each and every spontaneously generated animal as an instantiation of its species in the present, is a necessary part of the perpetual maintenance of that species. And that, for Liceti, is the highest end:

> concedere non possumus rerum imperfectiorum species in eo a perfectioribus diffidere, quod illae solum obtineant finem communem, qui est universi decus, atque perfectio; hae vero insuper et proprium; nam omnia, praeter finem communem, habent etiam proprium seu perfectioris, seu sint imperfectioris conditionis.[74]

> We cannot concede that a species of less perfect things differs in this [matter] from more perfect species on the grounds that the less perfect

> possess only the common end, which is the glory of the world and its perfection, whereas the more perfect [possess] a higher and proper [end]; for besides the common end, everything—whether it is in a perfect or imperfect state—has a proper [end].

And so the final cause of a spontaneously generated gnat, when considered with respect to what is proper to the individual gnat, is, as we saw in the passage just above this, simply the existence of the gnat itself, but in a larger, common sense, it is the perpetuation of its species for the glory of the world.

Liceti thus stands as the synthesis of the two dominant traditions of understanding spontaneous generation that we have seen so far, wedding Aristotle's mechanics and larger explanatory framework to Augustine's Christian creationism and something like its protogenic seminal principles. For Liceti, these seminal principles are perhaps a little more obviously material than they may have been for Augustine, and they actually occupy measurable space (as in a container), but we can see that the theological need for them is traceable to the same instincts. Moving forward in time from Liceti, we see that, as his hyperscholastic method of argument and scientific explanation goes out of style over the course of the seventeenth and eighteenth centuries, the Aristotelian packaging of Liceti's theory will be discarded, but the idea that the souls of spontaneously generated animals were all made in the beginning by God will come to dominate, and it will do so across some radically different theories of matter and generation.

Interlude: Is Life Special?

A GREAT DEAL of what has been so far driving our theories about how spontaneous generation happens, and a great deal of what will drive the opposition to spontaneous generation in chapter 6, is the (often gut-level) question of how special a thing life is taken to be. Is life so very different from raw, nonliving matter? Is it such a unique and miraculous phenomenon that it cannot be explained as originating by strictly material processes? If life is something that transcends strict (or even pregnant)[1] materiality, then questions arise about what causes new living things to come into existence, how the matter that animals are made up of becomes animated, structured, and sentient. Perhaps life can only come to be from other life (sexually and/or through decay), or perhaps the creation of a new life requires the intervention of a deity (either once in the beginning or continuously in the present).

The alternative to these positions is to argue that life is in fact not something so very special after all, that it can come into existence through simple material processes of one sort or another. We have seen Lucretius stake out such a position already: life arises when a particular combination of atoms becomes entwined in just the right kind of way. Descartes would famously pick up a similar mechanistic thread in the sixteenth century, although without Lucretius's concomitant atomism and with a very different conception of the human soul.[2]

So either life requires only a material explanation, or life requires more than a material explanation. We should note that these two positions are both—equally—external to the actual observations. We can see this quite clearly by considering for a moment an argument made by Henry

Charlton Bastian in the pages of *Nature* as late as 1870—long, long after Redi and Spallanzani, and even after Pasteur's experiments. Both a materialist and a proponent of spontaneous generation, Bastian pointed out that for opponents of spontaneous generation,

> a living thing has been supposed to be a something altogether different, incapable of arising out of a mere collocation of matter and of motion; and, therefore, under the influence of this theoretical assumption, whilst chemists and physicists have thought that they could in a measure account for the genesis of crystals by reference to the affinities and atomic polarities of the ultimate constituents of such crystals, [my opponents] have, for the most part, declined to account for the appearance of the minutest living specks in solutions containing organic matter . . . In the case of living things . . . the doctrine *omne vivum ex vivo* has become almost one of the "forms of thought."[3]

"Doctrine," "assumption"; he even goes on to call the idea that life supervenes on mere materiality a "theoretical view." Bastian makes it abundantly clear that what opponents of spontaneous generation all share—and this is particularly significant in the messy experimental environment of the debates in the 1860s and '70s—is a belief that life is so special that it could never arise from nonliving matter (as though life were a heritable, and only a heritable, trait). Any experiments where life shows up "spontaneously" must have, on this view, hidden methodological errors. Just look at the wording: "[my opponents] have . . . declined to account for the appearance" of life in organic solutions. It's not, for Bastian, that they have proved that life will never come into existence from nonlife. It is that, blinded by prejudice, they simply refuse to even try to explain what is happening in their vials in more than a hand-waving kind of way.

At the same time, although he was in the materialist minority, Bastian did not accept that his own position was a mere starting assumption, a choice external to the still-ambiguous experimental record. And here it is worth keeping in mind that Bastian would both win and lose this particular argument. True, we no longer believe in the spontaneous generation that he thought he was witnessing under the microscope, but at the same time, we do agree with him on the larger-picture claim that life can come to be from nonlife. We may believe life from nonlife to be a rarer phenomenon than Bastian did, but all the same, the origin of life in the universe is a question with significantly more tentacles, as it were, or significantly greater reach, than what was happening in one man's test tubes in 1870. Something like his way of thinking about matter and life has survived, even if some of the observations that added to his body of proof for that way of thinking, have not.

Turning back to our sources for the foregoing chapters, it appears that Aristotle strikes a kind of intermediary position between the two extremes. On the one hand, he clearly did not see much difference between living and nonliving matter. On the other, his theory of spontaneous generation requires the omnipresence of pneuma, something he explicitly calls "more divine" than the four elements, to account for the coming-to-be of life from inert matter. Surely this is more than materialism. But one is tempted at the same time not quite to call it vitalism either—after all, it's not that matter itself is fundamentally different from living things, it's that matter is just all potentially living in some sense. And it is quasi-divine—the problem of classifying this position becomes quickly apparent.

That living matter is not particularly different from inert matter for Aristotle can be deduced from several passages in the *Generation of Animals*, where he sees nonliving matter as full of vital pneuma (and so therefore always potentially living), but he also more explicitly draws our attention to this point in a passage of the *Parts of Animals* where he is discussing sea squirts and sponges:

> τὰ δὲ τήθυα μικρὸν τῶν φυτῶν διαφέρει τὴν φύσιν, ὅμως δὲ ζωτικώτερα τῶν σπόγγων· οὗτοι γὰρ πάμπαν ἔχουσι φυτοῦ δύναμιν. ἡ γὰρ φύσις μεταβαίνει συνεχῶς ἀπὸ τῶν ἀψύχων εἰς τὰ ζῷα διὰ τῶν ζώντων μὲν οὐκ ὄντων δὲ ζῴων, οὕτως ὥστε δοκεῖν πάμπαν μικρὸν διαφέρειν θατέρου θάτερον τῷ σύνεγγυς ἀλλήλοις.[4]
>
> Sea squirts differ little from plants in their nature but they are more animal-like than sponges are, for those have the nature of plants entirely. Nature transitions smoothly from lifeless things to animals, through things that live without being animals, such that there seems to be the tiniest distinction between one [kind] and the other in their nearness to each other.

He lays this position out even more clearly in the *History of Animals*, where he says,

> οὕτω δ' ἐκ τῶν ἀψύχων εἰς τὰ ζῷα μεταβαίνει κατὰ μικρὸν ἡ φύσις, ὥστε τῇ συνεχείᾳ λανθάνειν τὸ μεθόριον αὐτῶν καὶ τὸ μέσον ποτέρων ἐστίν. μετὰ γὰρ τὸ τῶν ἀψύχων γένος τὸ τῶν φυτῶν πρῶτόν ἐστιν· καὶ τούτων ἕτερον πρὸς ἕτερον διαφέρει τῷ μᾶλλον δοκεῖν μετέχειν ζωῆς, ὅλον δὲ τὸ γένος πρὸς μὲν τἆλλα σώματα φαίνεται σχεδὸν ὥσπερ ἔμψυχον, πρὸς δὲ τὸ τῶν ζῴων ἄψυχον.[5]

> Thus nature changes in tiny increments from nonliving things to animals, such that in the continuum the boundary between them escapes us, and of each there is an intermediate. After the genus of the nonliving comes first that of plants, and of these one differs from another particularly in how [much] it seems to have a share of life. But the whole genus [of plants] appears almost lively compared to other matter, even if [plants seem] lifeless compared to animals.

But notice the wording: plants appear "almost lively," σχεδὸν ὥσπερ ἔμψυχον,[6] compared to nonliving matter. Aristotle is here putting his finger directly on the essential tension in his theory: on the one hand, any foamy matter just lying in the sun seems to have the ability to produce life, and, on the other hand, living things, even the lowliest plants, have properties of growth, development, and self-organization that set them quite apart from nonliving matter. The contrast is meant to be a strong one, given the parallel clause at the end of the passage, where plants are "lifeless" compared to animals.

For Aristotle, then, there is something that sets even the simplest living matter apart from inert matter, so much so that we can talk about a sponge—*a sponge*—as sprightly by comparison with a rock. But at the same time, given his theories of matter and of generation, life is not particularly hard to create. Recall a passage we saw in an earlier chapter: γίγνονται δ' ἐν γῇ καὶ ἐν ὑγρῷ τὰ ζῷα καὶ τὰ φυτὰ διὰ τὸ ἐν γῇ μὲν ὕδωρ ὑπάρχειν, ἐν δ' ὕδατι πνεῦμα, ἐν δὲ τούτῳ παντὶ θερμότητα ψυχικήν, ὥστε τρόπον τινὰ πάντα ψυχῆς εἶναι πλήρη, "animals and plants are generated in the earth and in water because there is water in earth, and in water there is pneuma, and in all pneuma there is soul-heat, such that, in a certain sense, everything is full of soul."[7] Given the phrasing here and the way he talks about pneuma elsewhere in his biological writings, I take Aristotle to mean in this passage that pneuma is dispersed throughout matter in some way that is more or less all-pervasive. There is some ambiguity about whether, if pushed, Aristotle would say that the pneuma is in the air in tiny, distinct pockets ("as in a container," *à la* Liceti), or whether the two substances are truly "mixed,"[8] such that pneumated air is only potentially, but no longer actually, distinct bits of pneuma contained in distinct bits of air (in this case Aristotle would say that they remain separable in potential, even if no longer actually separated in practice).

We may be tempted by another familiar passage to think that he does not mean to say that pneuma and air form a completely blended mixture, since, as I remarked earlier, he seems to rather pointedly avoid the use of the word *mixture* when talking about the relationship between pneuma, water, and seed: ἔστι μὲν οὖν τὸ σπέρμα κοινὸν πνεύματος καὶ ὕδατος, τὸ δὲ πνεῦμά ἐστι θερμὸς ἀήρ, "seed is a sharing of pneuma and water, and pneuma is hot air."[9] I suspect in this instance he likely was doing so because he thinks of semen as a foam, literally a liquid with bubbles in it, and so is emphasizing that aspect of the material. But that does not help us with the present question about how soul-heat/pneuma and air (not water) are combined for Aristotle, and in fact this last passage complicates things, because now, instead of saying that there is pneuma in all air, as he did a moment ago, he is saying that pneuma is just air with a certain property: heat. I, for one, cannot see how to resolve the ambiguity. Whatever Aristotle's definition of pneuma is, though, and however he thinks it relates to air precisely, it is clearly not the same thing as Liceti's God-created souls-in-containers. Even if we take the pneuma to be a bubble in the surrounding air, or the pneumated air to be a bubble in the surrounding water, these are not the same thing as saying that each bubble is a single, finite, and individuated soul, which is what Liceti was arguing for.

Whatever way Aristotle thinks that pneuma inheres in matter, whether as discrete bubbles or as some subtler interpenetration, it is clear that he thinks it is a fairly trivial affair for pneumated matter to turn into living things. Pneuma is, at the same time, a stuff that has properties beyond the four elements of earth, air, water, and fire, properties that, as we have seen, Aristotle likens to the ether of the stars and planets. It is, to use his wording, "more divine." Thus, once plants and animals come to be living things, once they are separated off from their warm, frothing surroundings by a skin of some sort, these new entities exhibit remarkable, goal-directed behaviors that no thing-without-a-soul can match. Life, it would seem, is at the same time both remarkable, and remarkably cheap.

Not so for Augustine and Liceti, who require something more than the mere properties of matter for life. But what about those interim Aristotelians: Averroes, Albertus, and Aquinas, for whom the stars acted as efficient causes to generate life? Here, as with Aristotle's quasi-divine pneuma, much will depend on how the stars are supposed to be acting. If their influence is material, as it is for Ptolemy and many other astrologers, then life comes to be from material processes. If their influence is divine or quasi-divine, then we start moving into nonmaterialist readings. We might also ask what a soul-in-potential, inhering in matter down here on earth, might really be. Is it *material,* strictly speaking?

At this point, the problem threatens to overwhelm the current project. For the moment, though, I simply want to float the question of the transcendence versus the materiality of life, and with it the related question of just how special life as a phenomenon is. As we move into the final chapter of this study, we will see that this same question will underscore one of the great final debates in the history of spontaneous generation. And the positions taken up by the various parties as they pointed at their test tubes and argued through—rather than with—one another, were no more and no less rooted in the actual experimental evidence than were those of the thinkers briefly surveyed here.

CHAPTER SIX

Toward a Showdown

At the opening of this book, I called spontaneous generation a fact rather than a theory, on the grounds that for most of its history the spontaneous generation of various animals was really a set of raw data. The theories (plural) that then explained how spontaneous generation could work needed to be fit to that data. There was of course some question about how each kind of animal was generated: had it ever been seen to mate? Was it ever found with eggs or live young inside? Could we actually see it emerging, as with testacea from simple slime or, as with eels, from earthworms? With very few exceptions, the spontaneous generation of certain animals was no more or less a fact than the sexual generation of other animals was.

In the early modern period, however, this situation begins to change dramatically. Much of the context has shifted. The near universality of the idea that generation is the process of the imposition or the coming-to-be of a form in matter in the first instance was seriously waning, and legitimate alternatives to the Aristotelian explanatory edifice were floated with plausibility. Atomism was posing the first serious challenges to teleological philosophy in many centuries. Mechanism worked at an explanatory level that—if you accepted its premises—made talk of forms and of potential look vague, hand-waving, and disingenuously slippery. At the same time, some of the new approaches posed rather serious theological issues. For all that Descartes thought that God still underpinned his entire mechanistic system, for example, many of his readers thought that mechanism threatened or even implied a thoroughgoing atheism. Ralph Cudworth, for instance, in his 1678 *True Intellectual System of the Universe,* objected that

Descartes' God existed with no loftier purpose than to lurk in the background and watch things happen: "those Theists, who Philosophize after this manner, by resolving all the Corporeal *Phenomena* into *Fortuitous Mechanism,* or the *Necessary and Unguided Motion of Matter,* make God to be nothing else in the World, but an *Idle Spectator* of the Various Results of the *Fortuitous* and *Necessary Motions* of Bodies; and render his Wisdom altogether Useless and Insignificant, as being a thing wholly Inclosed and shut up within his own breast, and not at all acting abroad upon any thing without him."[1] As an alternative, Cudworth proposed that God had created the universe such that it was permeated by an omnipresent power or force that guaranteed order and regularity in the world. He called this power the *plastic nature.*[2] It was, he said, subordinate to God and acted as the instrument that effected divine providence through the laws of nature. It had no will of its own, but simply, "drudgingly," caused the order that we see in the universe. It could always be suspended or interrupted by God, who in any case always "presided" over it, which meant, to Cudworth, that "by this means the Wisdom of God will not be shut up nor concluded wholly within his own Breast, but will display it self abroad, and print its Stamps and Signatures every where throughout the World."[3] And it was animal generation that furnished one of the central proofs, for Cudworth, of the existence of this plastic nature:

> Unless there be such a thing admitted as a Plastick Nature, that acts ἕνεκα τοῦ, *for the sake of something,* and *in order to Ends,* Regularly, Artificially and Methodically, it seems that one or other of these Two Things must be concluded, That Either in the Efformation and Organization of the Bodies of Animals, as well as the other Phenomena, every thing comes to pass *Fortuitously,* and happens to be as it is, without the Guidance and Direction of any *Mind,* or *Understanding;* Or else, that God himself doth all *Immediately,* and as it were with his own Hands, Form the Body of every Gnat and Fly, Insect and Mite, as of other Animals in Generations, all whose Members have so much of Contrivance in them, that *Galen* professed he could never enough admire that Artifice which was in the Leg of a Fly, (and yet he would have admired the Wisdom of Nature more, had he been but acquainted with the Use of Microscopes.)[4]

Notice here the emphasis on and complete rejection of chance and the gestures at the regularity of nature: the insect happens to be as it is, with so much contrivance. He talks a little later about the "Regularity and Constancy every where."

More pointedly for our purposes, notice the emphasis on animals normally thought to be generated spontaneously, as well as the mention of a new instrument, one that is about to become the primary tool for investigating spontaneous generation: the microscope.[5] Part of the story of spontaneous generation is that for the last two centuries of its existence as a phenomenon, it was thought to be happening only or primarily at a microscopic level. The microscope provided not only evidence that tiny animals could be brought into existence in a controlled environment but also that those tiny animals did not seem to lay eggs, or not visible ones at any rate. We will return to the importance of the controlled environment of the test tube and sealed vial presently, but first we need to look briefly at the disappearance of macroscopic spontaneous generation, which happened in the wake of the work of Francesco Redi.

Redi's Fomentations

It is sometimes bemoaned that so excellent and careful an experimenter as Redi, one with his eyes so wide open to manifest evidence, should have failed in the end to prove—or even to believe it likely—that gallflies and other insects were *not* spontaneously generated in trees and fruits.[6] This is, after all, the man who performed one carefully controlled experiment after another to show that in case after case after case—meat, "stinking, rotten, and corrupted," fish, "completely changed into a thick and turbid liquid," eels "swelling and seething and slowly losing their shape"[7]—in all these cases no maggots were generated when flies were kept away from the rotting material. Spontaneous generation, it should seem, was disproved spectacularly.

But here is the rub: all these cases share for Redi one important feature: they would have, if they had actually produced life spontaneously, represented the generation of living things in dead matter. We have seen that it had long been held that insects and other animals could be generated spontaneously from both living and dead organic matter, and it is only the latter vector that Redi thought he had disproved. When Redi comes to consider the question of the spontaneous generation of animals in living matter, he changes his tune without missing a beat. Although he does briefly consider the possibility that these, too, might be the products of normal animal generation, which is to say of flies laying eggs, he seems almost dismissive. He "confesses" his temptation to believe that gallflies were generated from eggs laid by females,[8] and then he discards it in almost the same breath:

> Ma avendo poi meglio considerato, che vi son molti frutti, e legumi, che nascono coperti, e difesi da' loro invogli, o baccelletti, e che pur bacano, ed intonchiano; avendo osservato, che tutte le gallozzole nascon sempre costantemente in una determinata parte de' rami, e sempre ne' rami novelli; e che quelle gallozzoline, che nascono nelle foglie della quercia, della farnia, e del cerro anch' esse costantemente nascon tutte su le fibre, o nervi di esse foglie, . . . avendo ancora posto mente, che molte foglie d' altri alberi, su le quali nascono, o vesciche, o borse, o increspature, o gonfietti, pieni di vermi, quando quelle foglie spuntano, elle spuntano con quelle stesse vesciche, o borse, le quali molto bene si veggiono, ancorchè minutissime sieno le foglie, e vanno crescendo al crescere di esse foglie; . . . in oltre il cerro fa alcuni grappoletti di fiori; da que' fiori son prodotte altrettante coccole rosse, o paonazze, ciaschedduna delle quali ingenera tre, o quattro bachi rinchiusi ne' loro casellini distinti. Il medesimo cerro fa un' altro grappoletto di fiori, e da que' fiori spuntano alcuni calicetti verdegialli legnosi nella base, e teneri nell' orlo, e tutti questi calici fanno i lor bachi, ed i bachi escon fuora in forma d' animali volanti.[9]

> But then I thought that there are many fruits and pulses that come into being covered and protected by their rinds or their pods, but are still wormy and infested; and I observed also that all the growths consistently arise in a certain part of the branch, and always in young branches, and that those growths that are born on the leaves of the common oak, the English oak, and the Austrian oak always arise on the fibers or the veins of the leaves . . . and I further called to mind that many leaves of other trees on which blisters, bumps, swellings, or inflammations, full of worms, arise, when these leaves sprout, they sprout

> with these same blisters or bumps which can be seen no matter how small the leaves, and they grow as the leaves grow . . . To take another example, the Austrian oak grows some bunches of flowers and from these flowers are produced the same number of red or purple fruits, each of which generates three or four worms all shut up in their separate chambers. This same oak produces another bunch of flowers and from these flowers sprout little green growths, woody at the base and thinner at the tip, and all of these growths produce their worms, and these worms come out in the form of flying animals.

There seem to be three main observations driving Redi to spontaneous generation in these instances. One is that the excrescences from which the animals emerge seem to be there as soon as the leaf comes into bud, and they grow with (and therefore according to the same processes as) the leaf. Secondly, animals are born in fruits and legumes like beans and peas, which should have been protected from egg-laying by their skins. Finally, the regularity of the correspondence between plant species and insect species seems to point to generation *by* the plant rather than mere generation *in* the plant: oak gallflies are peculiar to oak trees in the same way as oak flowers are particular to oak trees (recall again the work that bees and cattle, wasps and horses did for so long in this story).[10] This contrasts sharply with the animals that Redi has observed born from putrefying matter, where the same species can come from different material and many species from a single festering specimen. This distinction between generation from living flesh and generation in inert, dead matter can also be seen in his discussion of mushrooms, where he tells us that solutions of macerated and rotting fungi are especially suited as a breeding ground for flies and other insects:

> io parlo però di que' funghi, i quali di già sono stati colti, e per così dire son morti, e putrefatti; imperocchè quegli, che stanno radicati in terra, o su gli alberi, e che vivono, sogliono generare cert' altre maniere di bachi, alcune delle quali sono differentissime nella figura in tutto, e per tutto da' vermi delle mosche.[11]

> I speak, however about mushrooms that have been picked and are therefore, so to speak, dead and putrefied. But those which are rooted

> in the earth or in trees and which are alive generate other kinds of insects, some of which are completely different in form throughout, and in every way, from the worms of flies.

He also later points out that if the gall is removed from the living oak prematurely, thus cutting off the "vital stimulus" that the developing animal receives from the tree, any insects within it will not live. Add all of this to the physical evidence of buds and veins and fruit rinds, and Redi finds it perfectly acceptable for some flies to be generated spontaneously in living matter. He attributes the mechanism to a principle of growth responsible not only for a plant's own development but also for its production of worms: *quell' anima, o quella virtù, la quale genera i fiori, ed i frutti nelle piante viventi, sia quella stessa, che generi ancora i bachi di esse piante,* "that soul or that principle which creates the flowers and the fruits of living plants, is the same one that produces also the worms of these plants."[12] This idea stirs in Redi a feeling of wonder and admiration for Nature, which he talks about repeatedly (he comments, for example, on the "marvelous craftsmanship," *la maravigliosa maestria*, by which nature has fashioned the vegetable mechanism for their generation).[13] Redi's conclusion is doubly remarkable, in that not even the great Aristotelian defender of spontaneous generation, Fortunio Liceti, believed that insects were actually engendered by trees and plants, believing instead that the plants picked up the germs of the insects from the water and soil, simply giving them a suitable home in which to germinate.

Redi's disproof of spontaneous generation in rotting matter was very well received, so much so that most naturalists after him went even further, to dismiss also his belief in the spontaneous generation of gallflies and other insects in living matter. Two oft-quoted passages from reference works will give the flavor of the mood shortly after his time.

In 1704, John Harris in his *Lexicon technicum*, under "Equivocal Generation," says that "the Learned World begins now to be satisfied that there is nothing like this in Nature; and since the Use of Microscopes, and a more particular Application to Enquiries of this kind, a prodigious Number of Plants have been discovered to have Seeds; and of Animals (Insects) have been found to be produced Univocally, or in the ordinary way of Generation, which before were thought to be Equivocally produced."[14]

Similarly, Ephraim Chambers, in his 1728 *Cyclopaedia*, offers us the following:

> EQUIVOCAL *Generation,* is a method of producing Animals and Plants, not by the usual Way of Coition between Male and Female, but I know not what plastic Power, or Virtue in the Sun, etc. See GENERATION.
>
> Thus Insects, Maggots, Flies, Spiders, Frogs, &c. have usually been supposed to be produced by *Equivocal Generation,* i.e. by the Heat of the Sun warming, agitating and impregnating the Dust, Earth, Mud, and putrified Parts of Animals.
>
> This method of Generation, which we also call *spontaneous,* was commonly asserted and believed among the ancient Philosophers: But the Moderns, from more and better Observations, unanimously reject it, and hold that all Animals, nay and Vegetables too, are *Univocally* produced, that is, by Parent Animals, and Vegetables of the same Species and Denomination.
>
> 'Twere a Thing, one would imagine, sufficient to discredit the *Aristotelian,* or rather the *Egyptian Doctrine* of *Equivocal Generation,* to find Flies, Frogs, Lice, &c. to be Male and Female; and accordingly to engender, lay Eggs, &c.
>
> To imagine that any of those Creatures could be spontaneously produced, especially in so romantic a Manner, as in the Clouds, as they particularly thought the Frogs were, and that they dropped down in Showers of Rain, were, certainly, highly unphilosophical.[15]

This terminology, calling spontaneous generation equivocal, was common throughout the Middle Ages and into the early modern period. The phrase "equivocal generation" was not, at least at first, as straightforwardly pejorative as we see it to be in the passages above. If we look to the definition given to it by Paulo da Venezia in the early fifteenth century, we see it as a simple definition covering a statement of fact, aimed at cases where animals are generated, but not from parents of the same kind: *generatio . . . equivoca est illa qua generans et genitum sunt diversarum specierum, ut generatio muli, seu muris, seu muscae . . . per putrefactione, seu mediante semine diversorum in specie*, "equivocal generation is that by which the generator and the generated are of different species, as in the generation of a mule, or a mouse, or a fly . . . through putrefaction or with an 'intermediate' seed from [animals] different in species."[16] Contrasted with *generatio univoca*, "univocal generation" (which is to say "generation properly speaking") the terminology was not entirely value-neutral, but it did not necessarily have the taint that would see later authors struggling to distinguish their law-like theories for the mechanism of spontaneous generation from the randomness later implied in their critics' use of the phrase "equivocal generation."

We see the line between univocal and equivocal generation being carefully demarcated, for example, in the Epicurean revivalist Pierre Gassendi, who, writing about a decade before Redi, had wanted to forestall any claims that his system for spontaneous generation was in any way "equivocal":[17]

> ut postremum quidem tam ista, quam illa causam habeant revera univocam, hoc est internam, essentialem, et ad unum determinatam; sed usus tamen obtinuit, ut causa illorum dicatur univoca, istorum vero aequivoca, quod respici solum ad causam externam, apparentemque soleat, non ad internam, et occultam.[18]

> Finally, in fact, [spontaneously generated animals] certainly have as univocal a cause as those [generated sexually]. It is internal, essential, and fixed for one end, but common practice is such that the cause of the one is called univocal and of the other equivocal, because only the external and apparent cause is usually considered, not the internal and invisible.

What makes it univocal for Gassendi is the preexistence in matter of *animulae*, "little souls," created by God in the beginning and waiting around in matter for suitable conditions to combine in the right proportions with each other and with nonliving matter, and so generate living things according to regular and law-like processes. This is very similar to the theory we saw floated by Liceti in the seventeenth century and by Augustine before him, except that in Gassendi it is wedded to a completely different theory of matter, that of Epicurean atomism.

As we saw in chapter 2, Lucretius had been entirely and doggedly materialistic: the soul is made up of fine and light nonliving atoms that become temporarily entwined with other atoms to create a living body. But this kind of strict materialism was never especially popular, and even in antiquity the peculiar Epicurean theological position that came along as a concomitant—the argument that the gods were entirely disinterested in paltry human affairs—often seemed a little too atheistic for most tastes. In Gassendi's day, the charge would remain a live one, even if Gassendi himself believed atomism to be perfectly reconcilable with Christian creationism and scripture.

A Blossoming of Theories

Gassendi's soul-infested atomism shares a characteristic, as I have remarked, with Liceti and Augustine, which is the idea that God created at least the building blocks[19] for all life in the beginning and embedded them as small physical particles of some sort in the material of the created world. I have called such theories *protogenic*, from the Greek for "created-at-first."[20] I have deliberately modeled this neologism on the contrasting theory, epigenesis—the idea that generation represents the coming-to-be of an entirely new life from matter that was not in any sense previously living. I have been calling these theories protogenic to emphasize the crucial point that it is not just that the soul or the configuration of the body exists prior to conception—even in the preformation theories that we will see in a moment, the thing exists (strictly speaking) before it is conceived—it is that the soul or bodily configuration was created in the beginning by God. *Protogenesis* as a word not only contrasts nicely with its opposite, *epigenesis*, but also helps avoid unwieldy terms like *preexistencism* and *preexistencist*, which sometimes want to crop up under the nomenclature more commonly found in the modern literature.

Epigenesis as a concept now begins to form an important touchstone as we come into the early modern period, so the terminological shorthand becomes necessary. In fact, given the increasing entrenchment and elaboration of (sometimes subtly) different theories of generation between the sixteenth and eighteenth centuries, it is necessary to briefly survey the different positions and commitments entailed by a handful of new terms. Readers already familiar with the differences between *preformation* and *preexistence* (my *protogenesis*), *emboîtement* and *panspermia*, can comfortably skip this section.

At this point we are quite familiar with Aristotle's theory of sexual generation, in which the male seed imposes form on the female matter to create a new life. For all that this theory seems to cover all cases of sexual generation and inheritance, it was not widely accepted even in antiquity, and most thinkers adopted two-seed theories, where the female was contributing something formal as well. By the seventeenth century, as these theories began to be elaborated and debated in greater detail, questions about what exactly was happening between the two seeds to produce the fetus began to tessellate into a wide range of divergent positions. Theories of generation—approaches to the question of precisely how a new life came into existence—proliferated, often entangled with the closely related but conceptually distinct embryological question of how the parts of the fetus came to be formed and developed.

One prominent and widely influential move was made by William Harvey as part of his larger epistemological and empirical inquiry into medicine and biology. Harvey's anatomical researches on animals caused him to dismiss even the modified-Aristotelian two-seed theory:

> nos autem asserimus (ut ex dicendis constabit) omnia omnino animalia, etiam vivipara, atque hominem adeo ipsum ex ovo progigni; primosque eorum conceptus, e quibus foetus fiunt, ova quaedam esse; ut et semina plantarum omnium. ideoque non inepte ab Empedocle dicitur, *oviparum genus arboreum.* habet itaque historia ovi fusiorem contemplationem; quod ex ea generationis cuiuslibet modus elucescat.[21]

> We maintain (and will prove in what follows) that all animals entirely, even viviparous ones, and even so far as man himself, are generated from the egg; that the first conceptions from which fetuses derive, are eggs of some sort[22] as is the case with the seeds of all plants as well. And so "the egg-bearing genus of trees" was rightly spoken of by Empedocles. The description of the egg therefore has an extremely wide coverage, since from it the manner of every kind of generation will emerge.

For all that Harvey is in many ways himself an Aristotelian, his theory of generation turns Aristotle entirely on his head, shifting the priority in generation from the male seed to the egg; in fact, Harvey posits the existence of a homogenous egg prior to the part-by-part formation of the fetus or sapling in every kind of animal or plant. The egg itself is the product not of the mother's uterus, as late Aristotelian researchers into hen's eggs had claimed (Fabricius is Harvey's source for this), but of her soul, and the fertilized egg contains within it a soul.[23]

Harvey's account of the primacy of the egg would prove to be massively influential, so much so that, as the author of an epitome of Caspar Bartholin's *De ovariis mulierem* noted in 1679, *l'opinion de la formation de l'homme par le moyen des œufs, aussi bien que de tous les autres animaux est quelque chose de si commun à présent qu'il n'y a quasi point de nouveau Philosophe qui ne l'admette aujourd'huy,* "the theory of the formation of man and of all other animals by means of eggs is so common at present that there is almost no modern philosopher who does not admit it today."[24] To distinguish this theory, that eggs have generative primacy from its (soon to be rapidly proliferating) competitors, historians follow early modern thinkers in giving Harvey's idea the name of *ovism*. But even this theory quickly took on a number of different forms. For Harvey, the life and the formation of the embryo in the egg began at the moment of fertilization: an entirely new life, and the beginning of embryological development, both happening simultaneously at the instance of conception. For both theological and mechanistic reasons, not everyone who followed Harvey would agree that conception marked the creation of a new life—is it not God, after all, who created all life in the beginning? And how could the many parts of animal bodies, so manifestly perfect for their uses, so fine and precise, come to develop from all but undifferentiated egg-stuff? Perhaps, some people argued, even the parts had been preformed by God in the beginning, hiding somewhere in the egg (or, soon after Harvey, in the semen) in invisible miniature.

And so, generation theorists who sided with Harvey in arguing that the living organism is created at the moment of conception called themselves *epigenesists,* with Harvey himself espousing, as we have seen, the theory of epigenic ovism. But as I mentioned above, other thinkers were often motivated either by concerns about what precisely happened at fertilization, by embryological concerns about how the parts of the fetal body were formed, or else—and this was the real kicker—by the theological problems posed by a new soul apparently popping into existence like this. These thinkers would come to form the strong majority in rejecting epigenesis, arguing for one of two alternative positions. For those who espoused preformationism—in its narrowest sense a position about the formation and development of the embryo rather than about the origins of life—the future fetus was supposed to be physically formed, with all the animal's parts already in their later arrangements, in one or the other parent's generative organs at some point before conception.[25] Conception was then a stimulus of some kind for the growth and development of these parts (differently elaborated in different authors). This neatly explains how the parts of animals come to have the form they have, the dog's nose having been formed in miniature before conception and then having grown to its full size by a simple process of augmentation afterwards. Returning to protogenic theories, in which all life was created in the beginning by God, these could obviously be perfectly compatible with the idea that the embryo was itself physically preformed, but I am going to concentrate here on theories of generation rather than of embryology and so will leave questions about preformation aside as tangential to our main question.

One very influential version of protogenesis is called *ovist emboîtement.* On this theory, the egg from which a baby develops has existed from the beginning of time in the mother, who had existed from the beginning of time in one of her own mother's eggs, who had existed in an egg in her mother, eggs nesting inside eggs, all the way back to Eve in the garden.

But of course there are other mechanisms by which God could have put the souls he created into the universe. One such mechanism, which we have already caught glimpses of, goes back to Augustine. It saw God as scattering seeds of some kind throughout creation rather than one egg in another in another, as *emboîtement* had it. *Panspermatism*, as it was called, solved a few of the problems that *emboîtement* posed (it could explain spontaneous generation quite easily, for example) but created a number of others.

Finally, there was one more protogenic theory that enjoyed a brief vogue, one that would emerge for a time as the main competitor to ovist protogenecism. This is the theory of animalculist protogenesis, which so dramatically entered the fray after Leeuwenhoek's discovery of "little animals" (*animalcula*) in semen. Animalculists thought that instead of all organisms being stacked like Russian dolls inside of eggs, they were instead stacked like Russian dolls in the spermatozoa of the male animal.[26] Instead of solving the problems that some people had with emboîtement, this simply moved the locus of emboîtement from the mother to the father. Worse, it created problems of its own, as we shall see.

It will at this point be prudent to summarize the range of these (often) modular, or pick-and-mix, theories and to clarify the terminology before proceeding, as follows:

epigenesis: the living organism is an entirely new entity when it comes into existence
protogenesis: all past, present, and future life has existed since God's single act of creation
preformationism: the embryo or its parts are preformed, prior to conception
animalculism: stresses the primacy of semen; egg merely serves as food
ovism: stresses the primacy of eggs; semen merely serves to activate growth/life
emboîtement: all descendants have existed since creation, stacked inside one another
panspermatism: all souls or germs have existed since creation, scattered throughout the world

Needham's Experiments

Whichever combination of theories one adopted, whether epigenesis or protogenesis, ovism, animalculism, or preformation, it had different implications for how one approached spontaneous generation, and the fact of spontaneous generation could be wielded as a stick with which to beat up one or another of these positions. Although spontaneous generation had largely, if temporarily, fallen out of favor after Redi, there were still serious puzzles around the generation of parasites, as well as questions about where all those little moving animals that could be seen in microscopes came from. And it was the little animals in microscopes that would become the focus of one of the last great debates on spontaneous generation—and the last for our purposes here—that between John Turberville Needham and Lazzaro Spallanzani.

Needham's work on spontaneous generation and his defense of epigenesis began with discussions he had with Buffon in Paris in the late 1740s.[27] Needham already had some significant accomplishments under his belt by this point, having been elected a member of the Royal Society in 1747 on the basis, in significant part, of his work in microscopy. In the event, Buffon had come up with a unique and novel theory of generation, and he was trying to persuade Needham of its plausibility. Buffon argued that the so-called animalcules seen in many liquids were in fact little life-like mechanical *parts* that combined to make larger animals when generation occurred. Needham, convinced that at least some of these self-moving *mechanica* were complete—if very small—animals, wanted to see if he could distinguish which of the infusoria were Buffon's mechanical parts and which were whole, independent animals, so he started working on various kinds of infusions that might harbor animalcules. He attacked the problem with vigor. First he heated up four different mixtures, sealed their vials from the outside air with corks, and watched them for days until he saw motions. He then set to work with Buffon to make fifteen more vials, checking them daily for signs of life. After just over two weeks, he and Buffon observed some exceptionally small self-moving particles, which they both at first took to be Buffon's mechanical "organiz'd Bodies," which is to say, mechanical parts.[28] But Needham, still worried about the sterility of his infusions, and in order to absolutely eliminate the possibility that these moving particles had come from the air outside the vial, then engaged on his most careful and far-reaching experiments. In his wording:

> I took a Quantity of Mutton-Gravy hot from the Fire, and shut it up in a Phial, clos'd up with a Cork so well masticated, that my Precautions amounted to as much as if I had sealed my Phial hermetically. I thus effectually excluded the exterior Air, that it might not be said that my moving Bodies drew their origin from Insects, or Eggs floating in the Atmosphere. I would not instil any Water, lest, without giving it as intense a degree of Heat, it might be thought these Productions were convey'd through that Element . . . I neglected no Precaution, even as far as to heat violently in hot Ashes the Body of the Phial; that if any thing existed, even in that little Portion of Air which filled up the Neck, it might be destroy'd, and lose its productive Faculty. Nothing therefore could answer my Purpose of excluding every Objection, better than hot roast-Meat Gravy secur'd in this manner, and exposed for some days to the Summer-Heat.[29]

Note his precautions in carefully sealing and reheating his vials. He is also very happy with the solution itself, in this case the mutton gravy hot from the fire, and he goes so far as to call it the "pure unmix'd Quintessence, if I may so call it, of an animal Body"—a remarkable claim when one thinks about it: the juices that come from a formerly living animal, reduced by heat, are the "quintessence," the very purest extract, "of an animal body." Needham, controlling for all factors, also performed the like experiment on "three or four Scores of different Infusions of animal and vegetable Substances . . . the Phials clos'd or not clos'd, the Water previously boil'd or not boil'd, the Infusions permitted to teem, and then plac'd upon hot Ashes to destroy their Productions, or proceeding in their Vegetation without Intermission."[30] What he got in every single case without exception—sealed or unsealed, heated or unheated—was life. Teeming, swimming, frolicking life. He famously concluded from this that the heat and careful exclusion of air made no difference to his investigation, and in words that stir to the horror of many a later commentator, "after a little time, I neglected every Precaution of this kind, as plainly unnecessary."[31]

In addition to his concern about eggs from the air, that he also had protogenesis in his sights the whole time is clear from his description of his subsequent observations of spermatozoa, which convince him that the tails of the spermatozoa were not actually a true part of the "animal," but instead were "in Effect nothing more than long Filaments of the viscid seminal Substance which they necessarily trailed after them."[32] In fact, the true animals in the spermatic fluid were just the small bodies at the top of the "tail," which were, he thought, generated spontaneously just like those in mutton gravy. That spermatozoa were really spontaneously generated animalcules is given strong confirmation in his description of the time sequence in his observations. He seems to indicate, although he is a little vague on the details, that the "filaments" emerged only over some time: "We saw a small Portion of male *Semen* plac'd on the Microscope, first, as it were to develope and liquefy, then shoot out into long Filaments, ramify on every Side, these open and divide into moving Globules, and trailing after them something like long Tails . . . They were of various lengths in various Animals, and they, insensibly, by the continual progressive Motion of those Animals, grew shorter and shorter, till some of them appear'd without any at all, swimming equably in the Fluid."[33] This evolution over time, this emergence of organization in the beginning, the growth of the filaments, and then the disappearance of the tails entirely in some of the animals, which afterwards remained unharmed, was proof for Needham that the animals themselves were spontaneously generated and that the tails were entirely a by-product of the medium in which they moved.

So spermatic animals were no different from any other microscopic organism, and Needham chalked up the spontaneous emergence of the entire class to a vegetative power that resides "in all Substances, animal or vegetable" (here reiterating the old emphasis on previously living rather than inorganic matter). And it is here that Needham makes a nod to the "little machines" of his co-investigator, Buffon, saying that his experiments seem "to imply that there are among [the microscopic animals], or not at a very great Distance from them, such as are only mere Machines, without any true spontaneity."[34]

Needham's worries about protogenesis begin with the standard, par-for-the-course objections that an epigenecist of his day might lobby. A standard—if not the standard—objection to the main theories of animalculism of the day pointed up the massive difference between protogenic animalculism and protogenic ovism (*emboîtement* versions). As we have seen, in protogenic ovism, it was argued that the egg that would come to form the new fetus contained within it any eggs that would come to form the next generation, which contained within them the eggs that would come to form the generation after that, and so on. Apart from the question of the limits of the divisibility of matter, and assuming we can get our heads around ever smaller eggs within eggs within eggs all the way from Eve to the future end of human history, the theory is self-consistent and theologically sound. But if we now look to animalculist protogenecism, we immediately encounter a problem of wastage. As one could plainly see under a microscope, the male semen contained not one, but innumerable little animalculi. If we now suppose each animalcule to contain within itself all the possible future generations of its potential line, nested within each other, we encounter serious problems. Take the case of a male animal, a dog for example, who impregnates a female. Out of the millions of animalcules, each a Russian doll of other animalcules, only a handful, maybe five or six, will become viable puppies and all the remaining thousands or millions of animalcules in the dog's semen—as well as the billions of animalcules each of those contains within itself ad infinitum—all of these come to naught. Of the ones that do survive to reproduce, the males among the offspring will eventually be in a position to give viability to, again, a mere handful of their own millions upon millions of animalcules, with each generation wasting a near-infinity of divinely created beings just to give birth to a few. Like many of his contemporaries, Needham found this impossible to accept. It presupposes, he said, "so great a Waste of Millions of Entities," countless souls within souls. All but a handful of these "most wonderful Productions in Nature" must be wasted at conception on this theory.[35] Secondly, how, on either animalculist or ovist *emboîtement*, does the non-embryo-possessing parent come to influence the features of the offspring?

Needham also worries repeatedly about the sheer mechanics, if you will, of the infinitely nested *stamina* (his word for the units of generation and ensoulment nested inside each other, whether these be eggs or animalcules). Think, he says, about how much smaller each *stamen* must be as we look back in time from an animal's progenitor to that animal's progenitor and so on. If the highest-level egg or animalcule is microscopic or even invisible in the living animal now, how much smaller must it have been when it was contained in the microscopic animalcule or egg of its parent, and how much smaller again in the *stamen* of its parent's parent? If we now run this back through all the generations to the moment of creation, how inconceivably small can the parts of the future organism have been when they were made? The compounded ratio of microscopicity, if we can call it that, of egg within egg going back hundreds of generations becomes simply staggering, no matter what one's theory of matter. And how can something that was originally so inconceivably small eventually become a visible, macroscopic being?

> If we consider the extreme Tenuity, I may say mere Nothingness of one of these *Stamina,* in its first Origin, at the Distance of so many Ages; comparatively to any one Part, the smallest muscular Fibre, for instance, of an adult animal it is now said to constitute: how can we understand, that so minute a Filament could be developed, or in any sense serve as a *Substratum* to a Cylinder so solid, so massive, so comparatively immense? Could a Mountain be look'd upon as a Superstructure

> upon a Grain of Sand? Or the terraqueous Globe derive its present Dimensions from the Dilation of an Atom?"[36]

Moreover, if the original stamen somehow physically increases, growing in size for generation after generation until causing or forming this or that animal, what, he asks (invoking the specter of Descartes and of atheism), would prevent us from seeing this as a mere mechanical process?

On top of these objections, Needham was impressed by the recent discovery by Abraham Trembley of the regeneration of polyps. It turns out—and this causes a massive problem for protogenecists—that if you cut one of these tiny animals in two, each separate part regrows to become a whole, living, independent polyp. How on earth could one protogenically created animal become two? One or the other of the polyps must be a new genesis, must it not? In which case, why not both? Adding to these questions the commonly known regeneration of limbs in lobsters and starfish (on which he himself had worked earlier), Needham points out that since anybody can see these parts growing back anew even in macroscopic animals, why need we suppose the original part was created in the beginning? Was the replacement? A little later he cites some experiments published by the Paris Royal Academy that succeeded in inverting a tree so that it grew upside down, such that "the Branches have become Roots, and the Roots Branches; a *Phenomenon* totally inconsistent with vital, essential, and unalterable *Stamina*."[37] After all, if God created the tree in the beginning, surely he must have created the roots and the branches. But now we see that the roots God created can be made to become branches and vice versa. Surely what God created was merely potential for life in matter, rather than each bit of each living thing, created fixed and immutable for its office?

If all this were not enough to convince us, Needham believes he has experimental evidence that disproves protogenecism. He tells us that protogenecism is "universal" among his contemporaries, "that every Plant proceeds from its specific Seed, every Animal from an Egg, or something analogous, pre-existent in a Parent of the same kind." A little later he calls this "the strong Prejudice of nearly two learned Centuries."[38] "It has generally been thought by Naturalists, that microscopical Animalcules were generated from Eggs transported through the Air, or deposited by a Parent

Fly, invisible to the naked Eye, or even that assisted with Microscopes. Yet is it strange that no Naturalist should yet have seen them, if they are really so numerous, when their supposed Progeny is so various, and themselves must be thought to be so frequently gliding over the Surface of all stagnant Waters."[39] Needham's contemporaries had extended Redi's experiments to their logical conclusion and argued that any microscopic infusoria found in solutions must be analogous to the visible maggots found in Redi's open flasks, and so likewise must have come from eggs somehow deposited there rather than being *sui generis* animals in their own right. Given this, Needham emphasizes for his reader the very serious problem of empirical evidence: his opponents should have seen the eggs they suppose the microscopic animals to have been formed from—such eggs should rightly be all over the place—and yet no one has ever laid eyes on them. More to the point, Needham has experimental evidence that at least some microscopic animals are the full-grown form of their species, insofar as he had, in cooperation with James Sherwood, demonstrated to the Royal Society that the microscopic animal called the paste eel bears its young alive. The experiment was an exceptionally fine one, too: it involved isolating a single paste eel under a microscope and then dissecting it with a very sharp blade to show that it was full of live young and therefore at its "ultimate State of Perfection."[40] (This experiment would become a standard exercise in handbooks on microscopy into the nineteenth century.)[41]

But it gets even better. It turns out that Needham did not just find his paste eels in nature. Instead, he could prove that they were generated spontaneously, in "a Mass from the clearest Spring-water, and the purest Wheat-Flour, heated as intensely as the Composition will admit."[42] And watch his wording here: "heated as intensely as the composition will admit." As Spallanzani will soon go on to show, if one simply boiled the mixture for long enough and also made sure to heat the air in the container sufficiently and also kept the container perfectly sealed and uncontaminated, no paste eels or anything else would be generated—or would not be generated most of the time, at any rate. This is all well and good in hindsight, two and a half centuries later, but for Needham and much of his audience, heating the mixture sufficiently to kill any animalcules was clearly heating the mixture sufficiently to kill any animalcules. Later in the treatise, Needham refers to "the Precautions I took, that no supposed Germs might either be convey'd through the Air or the Water, or remain adhering to the Substances infus'd; I have often, for these Purposes, made use not only of hot Broth, immediately closed up in a Phial, but also of pure animal Substances, such as Urine, Blood, etc. with the same Success; and in these, I believe, no one will suppose that Germs, Eggs, or Spawn, are pre contain'd if Care is taken to close the Phials immediately."[43]

Another very important observation that Needham took from his experiments was that the variability of generated organisms was not random or even very great—again, recall bees from cattle and wasps from horses. Instead, there seemed to be a clear regularity inherent in the process, such that only certain animalcules were generated in certain media. This allowed Needham to distance his theory, as Gassendi before him had done, from any theory of equivocal generation. The laws governing spontaneous generation are for Needham perfectly accessible to careful, patient experiment and observation. Regularity of generation, specific kinds of animalcules in specific media, the consistent generation of smaller animals from larger—all are important parts of his understanding of spontaneous generation. There is nothing equivocal about it: "these Principles, however capable of differing Combinations, yet admit only of a limited Variation, and never deviate further than is consistent with univocal Generation."[44] Indeed, Needham felt that he could reliably predict the sequence of animals, one generation of infusoria dying off and a new one arising from its slime, gradually collecting at the bottom of the vial: "thus I had for the Subject of my Observations what I may call a microscopical Island, whose Plants and Animals soon become so familiar to me, that I knew every animal Species, and every individual Plant almost without any Danger of a Mistake."[45] Elsewhere he continues, "and even in the lowest Class of microscopical Animals, I can truly say, that I never yet observed any others than Productions specifically determined; the same Substances giving the same Plants and Animals, and in the same uniform Order and Descent."[46] There were laws being followed in the production of these animals, one from another from another, and no external influence could explain that.

Needham accounted for the generation of infusoria in sealed and heated mixtures by appealing to a universal vegetative power inhering in all living matter. He thought that he could actually see this force at work before his very eyes. As his mixtures began to develop and change with the passing days, he could see little filaments forming (analogous to the filaments he saw in the semen earlier), and "these Filaments would swell from an interior Force so active, and so productive, that even before they resolved into, or shed any moving Globules, they were perfect Zoophytes teeming with Life, and Self-moving."[47] Later in the treatise, he says that the vegetative force is proved by "ocular Demonstration, which any Naturalist might have."[48] Moreover, he also had observations of his animalcules exhausting themselves and falling to the bottoms of his flasks, "motionless, resolv'd again into gelatinous filamentous Substance, and [these] gave Zoophytes and Animals of a lesser Species,"[49] according to the rules outlined earlier. That is, the animalcules that had spontaneously generated had themselves putrefied into generic matter of some sort (notice the implied homogeneity and neutrality of his "Substance" in the singular), which then spontaneously generated the lesser animals. Spermatic animals, apparently, do the same. The process, he says, repeats to the limits of his ability to observe it. This happens because animal and vegetable materials "return by a slow Descent to one common Principle, the Source of all, a kind of universal Semen; whence its Atoms may return again, and ascend to a new life"[50] (note that Needham does not mean by atoms here what Lucretius and Gassendi did; he is simply using the word as a shorthand for corpuscles or particles). A plant or an animal's vegetative force does not die with the dissolution of the body; instead, the matter of the body returns to a generic life principle (recall his gravy-as-quintessence claim earlier). And it is this universal vegetative force, species independent, that gives birth to the new animalcules.

In the end, Needham sees his experiments and the spontaneous generation of animalcules from infusions, and later of smaller animalcules from the decomposition of these first animalcules—always in a logical chain from greatest to least—as a strong proof of the existence and beneficence of God: "nor indeed can there be a stronger Argument deriv'd from any System of Generation whatsoever, of an All-wise Being, All-powerful, and All-good, who gave to Nature its original Force, and now presides over it, than from the Consideration of an exuberating ductile Matter, actuated with a vegetative Force, limited, tho we know not its exact Bounds, in its specific Ascent and Descent, and expanding itself in Directions as certain and determinate, as the Motions of the Planets."[51] This is significant because

Needham knows that, by making the life principle material in the way he has, he will be vulnerable to claims of Cartesian-type materialism, with which he had no desire to be associated.

Needham would come to root his theory of spontaneous generation in an idiosyncratic metaphysics (one much indebted to Leibniz) that saw the properties of matter, even basic properties such as extension and solidity, as ultimately the effects of the interactions between matter and our senses.[52] The real properties of matter, the properties that caused our senses to perceive solidity, extension, heat, and other qualities, those were accessible to reason alone, and Needham used his experiments on generation as a key clue to what those real properties were. What mattered most for Needham here was that the successive generations he was seeing in his infusions—life first coming-to-be from the infusion itself, then new life coming-to-be from the decaying substance of those organisms, ever smaller, ever less organized and sophisticated—this process was caused by the interplay of two principles inherent in all matter: the active and the resistive. Even the action of the vegetative force that Needham used to explain spontaneous generation could itself be reduced to the interplay of action and resistance. New life coming-to-be was the temporary triumph of action over resistance, and its aging and death the opposite. It is at this level of analysis that we begin to see the distinctions between life and nonlife break down. For Needham, chemical reactions and conflagrations—all of them—could be explained by the same two principles interacting with each other. None of this was worked out in any clear way in the text we have been following so far (Needham's *Summary of Some Late Observations*, presented to the Royal Society in 1748), but once Spallanzani published his own work, with its sharp criticisms of Needham's experiments and conclusions, Needham felt he had to elaborate the theory, and so the latter half of his notes on Spallanzani's book, published by Needham in French in 1769, would be occupied with a long physical, theological, and experimental elaboration.

Spallanzani's Critique of Needham

Spallanzani seems to have been originally motivated simply to repeat the experiments that Needham had performed in his 1748 publication, but he eventually became disenchanted with some of the finer details of Needham's method. In the end, he published the results of a long series of careful experiments in 1765. There, his attack on Needham begins with the accusation that philosophers fall into error when they base very general conclusions on too few experiments. Citing twenty-five of his own experiments, he says that only one gave him animals. Having said that, though, he admits that he thought he had seen animals in nearly all of his infusions, but these turned out to be, he says, false leads, falling dead to the bottoms of his flasks after a little while. He now switches register to tell us that in investigating this phenomenon he performed an "infinite number" of other experiments (later, "many hundreds," *centinaja ben molte di esperimenti*) that yielded results that would move him firmly away from Needham's conclusions.[53] At first he had trouble getting life in his infusions, but the following spring, nearly all his vegetable infusions were teeming, even after his having previously boiled them for an hour and a half.

But how? Spallanzani argues that the problem is not with the infusions themselves—he floats the idea that boiling the liquids for a few minutes should be sufficient to kill any eggs (at one point he mentions the effect that boiling has on chicken's eggs, for example). The problem is, he tells us, with the air. He begins by reminding his reader that Boerhaave showed the air to be full of *particelle terrestri, acquee, sulfuree, metalliche, saline, e simili*, "particles: terrestrial, aqueous, sulfurous, metallic, saline, and so on." He adds to this, as though there were no question even in Boerhaave about the matter, that these particles, *vi alberghino ancora, e quà e là svolazzando s' attacchino a' corpi, in che per sorte s' imbattono, verissime uova d' insetti già fecondate*, "are always there, flitting here and there, attaching themselves to bodies on which by chance they encounter the fertilized eggs of insects."[54]

It is with that image in mind—of the air we breathe positively teeming with microscopic eggs—that he turns to Needham's experiments, and in particular to the mutton gravy experiment that we have already seen. Needham had, recall, claimed that his "precautions amounted to as much as if [he] had sealed [his] Phial hermetically."[55] This would be, if Needham was correct, *l' ultima prova, che ad esso servì qual suggello a perenne stabilimento del suo Sistema*, "the final proof, which serve[s] as the seal on the eternal foundation of his system."[56] Spallanzani is here allowing himself a little joke, for it is the faulty seal on Needham's vial that Spallanzani will claim to be the undoing of this "final proof."

Spallanzani did admit, though, that his experiments were not at first entirely consistent. In particular, when he boiled his infusions but let them cool before hermetically sealing them, after a few days, *in alcune gli animaluzzi non erano venuti a luce, ma lo eran bene in molt' altre*, "in some, little animals did not appear, but in many others they were plentiful."[57] Thinking, as he tells us, that it may have been exposure to atmospheric air that was allowing eggs into his vials, Spallanzani decided to hermetically seal—actually hermetically seal, not just "as good as" hermetically seal—his vials before boiling them. This involved heating each flask intensely and drawing out the neck of the softened glass until it narrowed and the melting glass sealed itself entirely. Spallanzani reports that when he took these precautions, in every case the vials showed *nessun vestigio . . . di movimento spontaneo*,[58] "not the slightest hint of spontaneous movement," unless he pricked the glass to allow outside air to enter.

Needham was right to be suspicious here. Spallanzani was doing two things to his infusions that Needham was not: he was boiling them for much longer (Needham argued that one logically needed to boil them for at most the length of time that one boiled chicken eggs or silk-worm cocoons), and he was putting the enclosed air under considerable pressure in the process. Indeed, the pressure was so great that Spallanzani included a warning for any readers who might like to try his experiments out for themselves:

> prima di narrar l' esito passar non debbo sotto silenzio una cosa . . . ed ella è di star lontan dalle bocce ermeticamente serrate, qualor risentono la veemenza del fuoco, correndo rischio in quel tempo, per l' aer chiuso e incarcerato, che allor dilata possentemente le molle elastiche, correndo rischio, io dico, di andar spezzate, e i rotti pezzi volar per aria, come a me stesso toccò di vedere più d' una volta, vigente il colmo dell' ignea forza.[59]
>
> Before explaining the outcome, [there is] one thing I must not pass over in silence . . . and that is to stand back from the hermetically sealed vials when they are affected by the violence of the fire; there is a risk at that time because of the enclosed and sealed air, which forcefully

> expands its elastic particles,[60] there is a risk, as I say, of [the vial] shattering and the broken shards flying through the air, as happened to me on more than one occasion while tending the height of an intense fire.

This candid nod to the changing elasticity of the air is, as Spallanzani recognized, precisely where Needham could get a hook in, as Needham would in fact do in his rebuttal. Attempting to forestall Needham's objection on this point, Spallanzani tried to argue that the change in pressure of the air in the heated vials should have the same effect as the dilation of the air in a *voto boileano*, a pneumatic engine, which he had tried, taking the pressure to "nine inches of mercury"[61] without harming the infusions' ability to produce life.

Needham's Response to Spallanzani

Needham had Spallanzani's *Saggio di osservazioni microscopiche* translated and published in French in 1769, under the title *Nouvelles recherches sur les êtres microscopiques*, to which Needham then appended a series of chapter-by-chapter critiques, very nearly as voluminous as the text of Spallanzani himself, as well as a series of other discourses. As soon as his commentary opens, one senses Needham's touchiness at being accused by Spallanzani of resurrecting the old Cambridge-Platonic plastic natures. In the first paragraph of his reply, Needham goes on the attack:

> Si la Puissance impériale étoit plus avantageuse aux Romains après leurs guerres civiles, que l'Anarchie républicaine, où ils se trouvoient réduits, Jules César a eu raison de préférer au nom odieux de Roi, celui d'Empereur. La multitude se laisse conduire par les mots, plutôt que par des choses; et en cela, les Philosophes deviennent Peuple fort souvent.[62]

> Since imperial power was more beneficial to the Romans after their civil wars than the Republican anarchy in which they had found themselves, Julius Caesar had good reason to prefer the title of "emperor" over that of "king." The masses allow themselves to be led by words more than by facts, and in this same way, Philosophers [also] all too often turn into peasants.

That Needham sees a strong distinction between his own vital power and the "plastic nature" of his Platonic predecessors is clear enough from his brief discussion of them in his earlier *Observations*, where he had compared the proponents of the plastic nature to superstitious medieval scholastics, who thought that the planets were pushed around by angels. He adds there that the plastic nature has been adequately discussed and dismissed by others and needs no rehashing by him.[63]

Another important objection was that Needham thought that Spallanzani's protogenecism was less theologically rich than his own theory. He saw the key difference between epigenesis and protogenesis as cleaving along a line between whether God in the beginning had created *causes* or whether he created mere, and less interesting, *effects*:

> Qu'importe, pour assurer à la Divinité son empire sur ce monde matériel, et pour exclure les prétendus effets du hasard qui n'a jamais existé que dans les têtes imbéciles de gens abreuvés des fables du Paganisme, ou suivant les rêveries des Matérialistes, si les germes des corps organiques existent depuis le commencement de ce monde, formés immédiatement par son Créateur, ou si les loix générales par lesquelles cet univers est gouverné sont tellement fixées sous le bon et le sage plaisir de Dieu, qu'un tel effet spécifique doit nécessairement être produit par une telle cause prédéterminée? La dispute, quant à la Morale, roule sur un simple mot, sçavoir si Dieu doit agir immédiatement pour exercer son empire souverain sur la cause ou sur son effet, avec cette différence que ceux qui peuvent étendre leur vue jusqu'aux causes mêmes générales pour fixer les principes de la nature, sont certainement plus philosophes que ceux qui font agir la Divinité pour chaque effet en particulier; c'est ainsi que nos ancêtres employoient le ministere des Anges pour faire rouler les corps célestes. *Nec Deus intersit nisi dignus vindice nodus.* Hor.[64]
>
> When it comes to assuring God his empire over this material world and excluding the false effects of chance (which never existed except in the idiotic heads of people drowned in the fables of paganism or else chasing the daydreams of the materialists), what does it matter whether the germs of organic bodies have existed since the beginning of the world, formed immediately by its Creator, or whether the general laws according to which this universe is governed were so fixed by the good and

> wise will of God that this specific effect must necessarily be produced by that predetermined cause? As far as morality goes, the dispute turns on a simple word, to know whether God must act immediately to exercise his sovereign power over the cause or over its effect, with this difference: that those who are able to expand their view up to the general causes themselves to establish the principles of nature are certainly more philosophical than those who make God act [to bring about] each particular effect—this is how our ancestors used the ministry of angels to move the heavenly bodies. *Let a god not interfere unless the obstacle demands a hero* (Horace).

Again with the angels moving the planets: Needham has little patience for this deus ex machina level of explanation for life. Notice that this is almost, but not quite, the same problem as the worry Ralph Cudworth had earlier floated, that without the plastic nature one would have to suppose that God *immediately* concerned himself with the creation of every gnat, which sounds as if Cudworth has as his target a claim that God is busy creating every individual insect in real time. Needham's target is a God who does not create life in real time like that, but who instead has performed the miracle once and has since left off. Needham sees himself by contrast pleading for a universe that is causally sophisticated enough, and its matter richly potent enough, that it can generate life on its own. This, he thinks, is a subtler, more beautiful theology than any alternative.

When it comes to defending his experiments against Spallanzani's critique, Needham cuts immediately to the weaknesses in Spallanzani's technique that we flagged earlier:

> M. *Spalanzani* [sic] commence par me reprocher . . . de n'avoir pas multiplié suffisamment, à son gré, les expériences sur les substances exposées à l'action du feu. Mais si je parois avoir manqué à cet égard, je crois être bien plus en droit de lui reprocher d'avoir donné dans l'excès contraire à force de vouloir remédier à ce prétendu inconvénient.[65]

> Mr. Spallanzani begins by reproaching me . . . for not having sufficiently, in his opinion, repeated my experiments on substances that have

been exposed to the action of fire. But if I seem to have failed in this regard, I believe that I have much more right to reproach him for having gone to the opposite extreme in wanting to remedy this alleged flaw.

Too much heat, for Needham, was clearly where Spallanzani had gone wrong, and he does not hold back on explaining why: *en effet à quoi bon, sous prétexte d'exterminer certaines germes qu'il croit s'attacher aux parois intérieurs d'un vase scellé hermétiquement, ou dans la substance même de la matiere renfermée*, "what good does it do, under the pretext of destroying certain germs that he believes attach themselves to the inside walls of a hermetically sealed flask or in the substance itself of the enclosed material"—and here I want to break for a moment to draw attention to just how significant his wording is: "certain germs that he believes attach themselves." It is easy, from our standpoint today, to immediately see this as wrong-footed, but of course we must remember that the question of spontaneous generation, the question of whether the air in fact did contain a multitude of the submicroscopic eggs of microscopic animals, floating all around us—that is precisely the ball in play here. And no one, not Spallanzani, not anyone, has seen these germs. All we know, in 1769, is that if you boil a hermetically sealed flask for a really long time, you don't (usually) get microscopic organisms. The open question, then, is what does that boiling do? We must remember that here there are two very real—and very different—possibilities.

And so let us allow Needham to continue:

> en effet à quoi bon . . . vouloir trop gêner une expérience, et la mettre à la torture, pour la forcer à déposer faux contre le témoignage positif de la nature, qui réclamera toûjours en faveur de l'épigenese? Un seul quart d'heure, quelques minutes de cuisson dans de l'eau bouillante, doivent suffire assurément, pour détruire tous les germes prétendus qui peuvent se trouver dans l'intérieur d'un vase fermé hermétiquement, sans pousser cette cuisson, comme il l'a fait, jusqu'à une heure, ou même une heure et demie d'une ébullition très forte.
>
> Selon ses propres expériences, il paroît évidemment que tous les êtres microscopiques, en général, qui naissent dans nos infusions, périssent dans un fluide qui passe tant soit peu le dégré d'une chaleur mo-

dérée, que les germes des vers à soie sont détruits par l'eau bouillante en très-peu de tems, et presqu'en les plongeant simplement ou les y laissant un instant; que l'ardeur même du soleil suffit pour dessécher, et détruire les œufs de tous les insectes généralement, et qu'enfin les germes mêmes les plus durs des végétaux deviennent inféconds, par la chaleur excessive du feu, dans quelques minutes ou de cuisson, ou de torréfaction.[66]

What good does it do . . . to want to so extremely hinder an experiment and to subject it to torture in order to force it to bear false witness, contrary to the clear testimony of nature which always speaks out in favor of epigenesis? A mere quarter of an hour, a few minutes of cooking in boiling water should certainly suffice to destroy any supposed germs that might be found in the interior of a hermetically sealed flask without extending this cooking, as he has done, all the way to an hour or even an hour and a half of very hard boiling.

According to his own experiments, it is clear that all the microscopic organisms in general that are born in our infusions perish in a liquid that experiences even a moderate amount of heat, that the germs of silkworms are destroyed by boiling water in much less time—almost as soon as they are simply immersed or kept in for an instant; that even the heat of the sun is enough to dry out and destroy the eggs of all insects generally and that, finally, the most hardy of vegetable germs are made sterile by the excessive heat of a fire after a few minutes of boiling or roasting.

Needham makes an important point here. Spallanzani has not given any reason why the eggs that he supposes infest the air should require so very, very much more violence for their killing than any other egg or grown organism known to humanity. We know what eggs are, Needham says, and they simply don't work that way. This is not trivial: for Spallanzani's proof to work we must accept the existence of some new kind of super-egg, one that behaves in ways that are radically different from any other known eggs. Worse, we know from experience that the larger an egg is, the longer it takes to boil. Why should the opposite be true for the tiniest figments of eggs that anyone has ever even imagined?

Here Ian Hacking's work on what he calls "the creation of phenomena" is instructive.[67] For Hacking we don't just experiment, we also create phenomena in laboratories (often enough, phenomena that happen only in laboratories). As we set up, manipulate, and tweak our apparatus, we often do so in order to affect how entities that we cannot see behave, and how the behavior of those entities then goes on to cause other effects: we use electrons in order to study something else, say. So what is happening between Needham and Spallanzani has much to do with which invisible entities each thinks that he (and the other) is really manipulating in his vials, and this deeply affects both the precautions and the positive actions that each sees as valid and truth-conducive. (We will return to Hacking in the conclusion.)

Remember, too, where Needham sees the real question as lying. He does not say that nature testifies in favor of spontaneous generation except when it is tortured by violent boiling; he says that nature testifies in favor of epigenesis, which is something else entirely. This matters because, whatever we may think is going on in their flasks, we—today—must concede that it does not come down on one side or the other of the actual question both Needham and Spallanzani thought they were putting to those flasks. For someone who believes in spontaneous generation, it is not difficult—as we've seen again and again—to believe that God created all beings in a single act at the beginning of the world. That marriage has been viable at least since Augustine. Hitching a theory like Spallanzani's ovist *emboîtement* to that idea could be straightforward, so long as one is willing to let God spread eggs throughout the cosmos, waiting for the right conditions to germinate. But that is not what Spallanzani does. Recall, for example, Spallanzani's careful wording when he describes how the air can pick up eggs from surfaces. It is not just eggs that the air picks up, but *uova d' insetti*, "the eggs of insects," and not just the eggs of insects, but *uova d' insetti già fecondate*, "the already-fertilized eggs of insects."[68] For him eggs were created in the beginning, but they also have only ever been contained in animals rather than all or some being spread throughout the universe by God as panspermaticists would have it. But that is an assumption and—strange as it may seem to say—it is unproved and unprovable by his experiments in the *Saggio*.

What Spallanzani sees as being at issue in the flasks is whether or not the nonliving matter itself may be turning into living organisms through the action of some force inhering in the infusion. Is it possible for life to come into existence from nonlife? If Needham is right, that this is what is happening when mutton gravy and other infusions fill up with microorganisms, then the birth of a new life must be a process of epigenesis. God cannot have created Needham's eels and other animals in the beginning, and that is exactly what Needham believed himself to have ruled out by boiling his gravy and heating his sealed vials in the coals.

And there is that question again—heating his vials for how long? In order to prevent the birth of microorganisms, Spallanzani had to resort to heating his vials violently enough and long enough that they sometimes exploded—think about that for a moment—on the assumption that the smallest eggs imaginable were the hardest eggs to kill. Is it not more likely, Needham argued, that such violence caused a critical change in the nature of the material encased in the vial?

> Voici un autre fait qu'on ne doit pas oublier et que je tiens également de lui-même: dans des vases ainsi scellés, la présence d'une certaine quantité d'air pur, proportionnée à la quantité du fluide contenu et de la substance macérée, est absolument nécessaire pour la génération de ces êtres vitaux dans le systême également des germipares, comme dans celui de l'épigenese: or, de la façon qu'il a traité, et mis à la torture ses dix-neuf infusions végétales sans aucune nécessité, s'il n'a cherché simplement qu'à détruire les germes prétendus qu'il suppose pouvoir exister sur les parois intérieurs de ses vases, il s'ensuit visiblement que non-seulement il a beaucoup affoibli, ou peut-être totalement anéanti la force végétatrice des substances infusées . . . mais aussi qu'il a entièrement corrompu, par les exhalaisons et par l'ardeur du feu, la petite portion d'air qui restoit dans la partie vuide de ses phioles; il n'est pas étonnant parconséquent que ses infusions, ainsi traitées, n'aient donné aucun signe de vie.[69]

> Here is another fact that we must not forget, and one that I take from him: in the flasks that are sealed in this way, the presence of a certain quantity of pure air, proportional to the quantity of the enclosed fluid

> and the macerated substance, is absolutely necessary for the generation of these living organisms, both in the system of the protogenecists and in that of epigenesis. Now, concerning the manner in which he handled and submitted to torture—with no reason at all—his twenty-nine vegetable infusions: while he had simply sought only to destroy the alleged germs that he supposes can exist on the inner surfaces of his flasks, it clearly follows that not only did he greatly weaken, or possibly even completely extinguish, the vegetative force of the infused substances . . . but also that he completely corrupted, with the exhalations and violence of the fire, the small portion of air that was in the empty part of his vials. It is not surprising, therefore, that his infusions, treated in this way, did not produce any signs of life.

Of course, in hindsight this looks plainly question-begging. Indeed, we can even recast it in the form of a syllogism:

(a) the vegetative force is responsible for the generation of microorganisms,
(b) long-term, violent heating prevents the generation of microorganisms, therefore
(c) long-term, violent heating destroys the vegetative force.

From the vantage point of the twenty-first century, one sees immediately where the problem is. But now let us write up Spallanzani's position in the same way.

(a) invisible super-eggs are responsible for the generation of microorganisms,
(b) long-term, violent heating prevents the generation of microorganisms, therefore
(c) long-term, violent heating destroys invisible super-eggs.

There is nothing—nothing—in either Needham's or Spallanzani's experiments that would allow us to decide between these two sets of propositions. We have exact explanatory symmetry, and, in 1769, the only criterion any reader of the *Nouvelles recherches* had to go on in order to choose between the two possibilities was the reader's own presuppositions about epigenesis versus protogenesis. This is particularly interesting given the way that the history of biology would develop after 1769. For, of course, we would eventually build microscopes that would allow us to see the previously invisible super-eggs. At the same time, though, we would also come to side with Needham on epigenesis.[70]

Which, I ask in lieu of a conclusion, would Needham and Spallanzani have seen as the more far-reaching of these two outcomes?

Conclusion

AT THE END of the last chapter I hinted that perhaps Needham was the real winner of the debate with Spallanzani—a strange note, historiographically speaking. This is not a claim I genuinely wish to endorse (it is not, in any case, the business of historians to pick winners and losers), but it does make a point, which is that from within the thick haze of a scientific controversy, it is impossible to really tell where the dust will eventually settle (watch that phrase "eventually settle": it's not nearly as innocent as it looks, as we'll see shortly). It is also not the case that the current scientific consensus came to epigenesis by going back and reexamining Needham's or Spallanzani's experiments. It's that in the fog of the debate over epigenesis versus protogenesis, most contemporaries thought that Spallanzani's disproof was a disproof of epigenesis first and foremost. But it only worked as a disproof of epigenesis in the experimental climate of the mid-eighteenth century, when epigenesis had become conflated with spontaneous generation through a series of historical and theoretical contingencies. The same set of experiments as Spallanzani's would have proved something very different to Augustine, for whom spontaneous generation happened because of, not in spite of, protogenesis. Indeed, Spallanzani's experiments are taken today to have proved something different; his work is hailed in biology textbooks as one of the great bulkheads in the war against vitalism and spontaneous generation.

I also stopped the discussion in the middle of everything. Spallanzani would go on to work very hard in responding to Needham's critique, and he would come back with a series of refined experiments and arguments in his *Opusculi di fisica animale e vegetabile* of 1776. I deliberately made the decision to stop short of this point in order to isolate a moment in the debate when the evidence was, from where we stand, still genuinely ambiguous, when the claims by both sides had symmetrical blind spots and symmetrical but perfectly opposed theories about matter, life, and the effects of heat. There is something telling about a moment when each side thinks it has won an argument, only to have the other side say that no, you actually contaminated your experiments by heating your vials too much, heating your vials too little.

Hitting the pause button at just this moment in time allows us to throw some light on a question in the philosophy of science that I have a particular interest in, the question of whether or how a belief in scientific realism is justified. Scientific realism is, to gloss it as generally as possible, the belief that the entities, models, or laws that the (best versions of) our sciences posit and use are plausibly interpreted as being real rather than convenient fictions or some such thing. There are a number of ways that the history of the sciences tends to complicate the defense of such a position. For example, holding a version of scientific realism in the past, in 1900, or 1800, or 100, would have seen us believing in the reality of a host of entities and forces that have not withstood the test of time. We would, in short, have been wrong to be realists in the past. So what, the objection runs, makes us right to be realists now?[1]

In the midst of the discussions around various aspects of scientific realism, there are currently a couple of live debates that may appear at first sight to be tangential but that cut pretty close to the heart of the issue. One stems from a provocative question about contingency and inevitability asked by Ian Hacking a little over a decade and a half ago, and the other involves a refinement of the old underdetermination-of-theory-by-evidence problem, sharpened by Kyle Stanford into the "problem of unconceived alternatives."[2]

Hacking's question was an apparently simple one: "Are the results of successful science," he asked, "inevitable?" Imagine if aliens or other intelligent beings were to have a science that was just as successful as our own—would they inevitably have the same laws, questions, and theoretical entities as we do? Would they have protons and DNA and inverse-square laws, or might there be another way of thinking about matter and life that could still be "successful" in the relevant sense but that would generate a different science from our own, one with different but just-as-effective laws, entities, forces, and so on? Much refinement has been brought to bear on this question in the fifteen or so years since Hacking first posed it, but the core issue is still a live one, and it sits at the heart of many philosophers' intuitions about realism versus its alternatives.[3]

A second wrinkle on this set of questions has been highlighted by Kyle Stanford, who asks about the alternative theories that might explain any given body of evidence. Ever since the work of Pierre Duhem, philosophers have been acutely aware that theories are underdetermined by the evidence that is adduced to support them.[4] That is to say, for any body of evidence, there is always in principle more than one theory that can adequately account for that evidence, and we know that historically, sometimes (often?) a much better theory has not even been dreamt of at a particular moment in time. This last point is what Stanford calls the problem of unimagined alternatives: at any given time, we cannot know how many alternatives—ones that we cannot even conceive of at the moment—there might be to explain a set of phenomena. No one in the pages of the present volume imagined DNA, RNA, or chromosomes. No one imagined bacterial spores, no one the big bang or the 13.81-billion-year span of development the universe has enjoyed since then. This seriously impedes their ability to weigh all the relevant alternatives for explaining what was happening in their test tubes, ponds, and manure piles.

Hacking is asking whether true, viable alternatives to our successful science really do exist, and Stanford is showing how, for realism, things becomes pretty thorny if so.

I want to offer my frozen moment between Needham and Spallanzani as shedding light on these important questions. Imagine that the debate between Needham and Spallanzani had played out differently after the 1769 encounter. Instead of Spallanzani continuing doggedly on and refining his experiments in ways that his contemporaries would find most convincing, instead of Needham leaving his own experimental program behind as being now unnecessary, believing himself to have made the point sufficiently—imagine instead that things went exactly the other way around. Spallanzani moved on to other work, but Needham refined his experiments carefully to show that overheating and deelasticizing the air and overboiling infusions was really the problem. It is not difficult to plausibly imagine him doing so and, with careful technique, silencing his critics and producing a triumphant proof of epigenesis while maintaining a belief in spontaneous generation.

So imagine that Needham had won. Suppose that subsequent science, for a time at least, instead of believing in the one now-right idea and the other now-wrong idea had it the other way around: believing instead in (now-wrong) spontaneous generation, but also (now-right) epigenesis. Assuming that science is progressive (in whatever sense the reader likes) one can still assume that history would have, within a generation or four, realized that the spontaneous-generation half of Needham's victory was in error, just as it eventually realized that the protogenesis half of Spallanzani's victory was in error. That only one half of the pair of phenomena, protogenesis and spontaneous generation, gets disproved on either scenario does not, it bears saying, make that subsequent science unsuccessful. It merely draws attention to the larger picture, which is, from where we stand now, unfinished, incomplete.

One question we can now ask is, on this alternate scenario, how would our science be different from what actually, historically happened? If the majority of practicing workers in biology had accepted, for that interim period, spontaneous generation while rejecting protogenesis, how would their practices have differed from what actually transpired? What research programs, what byways and highways would they have followed? What tools, instruments, and techniques would they have developed to solve the emergent problems of the combination of epigenesis and spontaneous generation? How would theories of matter have developed in a climate in which, for European science, God—from an earlier date than historically happened—did not have to intervene by planting the seeds of every insect at the moment of creation? And how would theories of life have developed in a late-eighteenth-century climate in which life itself was simply not all that special, in which apparently inert matter was imbued with a vegetative power? What discoveries and side effects would have turned up as research into the nature and scope of that power developed? In short, how different would science have looked when, eventually, biology settled out just as it did historically, believing neither in spontaneous generation nor protogenesis? Surely things would be importantly different, the tools, techniques, priorities, and even the perceptible or theoretical entities of subsequent biology having developed in different ways. This is a spin that I think may, for several reasons, add something to our thinking about unimagined alternatives and contingency versus inevitability.

As I have argued elsewhere, an important part of the point here is that we live—that we have always lived—on the crest of the historical wave, highlighted by Needham and Spallanzani where we left them hanging in 1769: the people who get to decide which science counts as successful, which experiments are reliable, which theoretical entities and processes are trustworthy, are always and absolutely forward-blind. Standing on the ground in 1769, no one—no one—could know what was coming in the future, how the various theories at play would eventually resolve. They could not know where the dust would eventually settle—which is precisely the catch that the contingency and inevitability debate runs into: like every single one of our forebears, we ourselves are also completely, utterly, forward-blind. Because what counts as successful science is to be adjudicated in every single case by thinkers who are forward-blind, by thinkers who have no idea what may still be true in fifty or a hundred or a thousand years, the question of the inevitability of the results of successful science runs up against a wall. 'Successful' as a criterion is inevitably contingent.

Here there is a standard strategy employed by a number of realists to try and corral the problem. It is all too easy for antirealists to point to historical examples of what now appear to be mistaken theories about the world, examples in which realist conclusions about psychic pneuma, N-rays, or caloric would have been premature. As a line of defense against this, many realists have adopted one or another strategy from a family of responses that tries to qualify either which sciences we can call genuinely successful or which parts of historical or contemporary sciences genuinely lead to their success. So we can accept Maxwell's field equations as genuinely successful and realism-conducive even if we now reject the luminiferous ether in which those equations were once embedded. Alternatively, we could adopt the popular strategy of cordoning off success and limiting it to sciences that have particular qualities: "maturity" and "predictive success" being longtime favorites, such that only mature, predictively successful scientific theories are genuinely, realism-justifyingly successful.

But here, forward-blindness rears its head again. Where previously it prevented our historical actors from knowing where the dust would settle in their own debates, here it threatens the very criteria we are invoking for adjudicating their debates as well as our own. We simply do not know what epigenesis will look like after the passing of more time. To ask if epigenesis is inevitable for any successful science is to assume that epigenesis is a finished product of science, no longer subject to question, significant refinement, rejection. If we attempt to quarantine this move by arguing that epigenesis is the result of mature, predictively successful science, then we are faced with the much more mature science that gave us spontaneous generation and protogenesis for so many centuries in the first place, a science that was—from where its practitioners stood—also eminently predictively successful.[5]

Think about it this way: one of the problems with asking the question of the inevitability of successful science is, clearly, finding the criteria for what will count as successful. What is so interesting about stopping the Spallanzani-Needham debate in 1769 is that it doesn't matter which of the two thinkers is seen by their contemporaries to have won—either result is, from where we stand now, a successful move in the right direction. And that is precisely the problem. It could have gone either way—either way would have been successful—but the particular direction it went in was entirely contingent. The inevitablist rejoinder here is to say that, while this may be true, we did go on from there to correct both parts of the equation (spontaneous generation and protogenesis), so although there may be a trivial contingency about the order of the events, it's still inevitable that our successful science ended up where it did for both phenomena. This is all well and good, perhaps, but it assumes that things now are more stable than they were in 1769, that protogenesis and spontaneous generation are done and dusted once and for all. Saying that is to invoke, as I've pointed out elsewhere, a kind of transhistoric privilege that we, and only we, enjoy. Although Spallanzani and his contemporaries thought protogenesis to have been proved, and to have been the (inevitable?) result of successful science, we today have the privilege of looking back and adjudicating that assessment as premature, forward-blind.

As I said above, though, we are just as forward-blind as Spallanzani was. We may think we have improved things enough, learned enough, fine-tuned our criteria enough to insulate us, now, from problems that plagued him, then. But he was once on the crest of the historical wave himself, and a philosopher's own criteria for what might at that time justify a realism were just as state-of-the-art as ours are (by which I mean that they were the most up-to-date criteria that had ever been), and yet they would have turned out to be misplaced and premature. Running the parallel thought experiment, in which Needham emerges the victor for a time, shows that science can be perfectly successful without being finished. Hacking's question about whether the results of successful science are inevitable or contingent turns out to be less about "successful" science and more about "finished" science—which raises the bigger question: what on earth would finished science look like?

These bigger questions are, of course, considerably more than this little book can hope to take on, but I think they are worth considering. And I think that Needham and Spallanzani, frozen in time as we've left them, have something to say going forward.

Introduction

1. For histories of spontaneous generation in the period covered by this book, see van der Lugt, 2004; Hirai, 2005; Farley, 1974; Gasking, 1967; Roger, 1963; Harris, 2002; Deichmann, 2012; von Lippmann, 1933.

2. See, e.g., Roger, 1963; Des Chene, 2000; Mendelsohn, 1976; Farley, 1974.

3. That being said, if we also include medicine, a good case could be made for the various etiologies covered by the name *hysteria* over roughly the same period.

4. See, e.g., Harris, 2002.

5. Mendelsohn, 1976, 40.

6. On the transmission of Aristotle to the later traditions, see, e.g., Brams, 2003; De Leemans, 2010.

7. The seminal study is Coady, 1992; see also Kusch, 2002.

8. On traveling facts, see Howlett and Morgan, 2010.

9. My use of the word *biomechanisms* is not intended to push my authors to one side or the other of the early modern debates around whether biological processes were mechanical or whether processes like generation involved some properties that supervened upon mere mechanism. I simply cannot think of a better word to describe the level of analysis at which authors are working as they try to explain, at the minutest level, what is physically happening when an animal is being generated. I also will talk generally about *physical* processes in generation, which is meant to include biological and chemical processes and, for some authors treated here, can include the interactions of souls.

10. See chs. 9 and 10 of Lehoux, 2012.

11. The genus of testacea (bivalves and snails) is the exception for Aristotle because he thinks the members are more like plants than other animals (I am unsure if he is primarily thinking about bivalves rather than snails in those passages). He also says that testacea are all generated spontaneously (making it look as

if that feature is generic, even if not limited to this genus), while at the same time he worries about what it means that snails have been seen mating.

12. Kuhn, 1962.

13. Of course, this stage is followed for Kuhn by "revolution," which establishes a new paradigm to get the whole cycle going again.

14. I readily recognize that my use of the term *biological* and its related terms (*biology, biologist,* etc.) are anachronistic, but I trust not perniciously so. There are contexts in which the use of such anachronisms really matters, and others in which, I think, they are more innocent. For the present work, while we should recognize that Aristotle did not, and could not, have seen his work as defined by the category that we today put under the umbrella of *scientific biology*, and that his priorities and ways of thinking were often very different from those of a modern biologist—recognizing all of this, I will go on in this work to refer to the group of Aristotle's work on animals and living things as his *biological corpus*, or even just his *biology*, as a simple shorthand to avoid wordy circumlocutions. Similarly, with terms like ancient *science*—I know the problems with using such terms too literally, so I do not use them in that way. Furthermore, when I do use them, I do so with all the caveats above in mind. For the issues at play here, see, e.g., Lloyd, 1970, 1979 (both of which in any case use terms like *science* liberally and without egregious anachronism); Cunningham, 1988, 1991, 2000; French and Cunningham, 1996; Cunningham and Williams, 1993; Cohen, 2010.

15. Van der Lugt, 2004; Hirai, 2005; Farley, 1974; Gasking, 1967; Roger, 1963.

16. What I mean by this way of talking and why I do so is explained in chs. 9 and 10 of Lehoux, 2012.

Chapter 1 · *Spontaneous Generation in Aristotle*

1. For my scare quotes around "scientific revolution," see Shapin, 1996.

2. Ebrey, 2015; Tipton, 2014; Leroi, 2014; Gotthelf, 2012; Leunissen, 2010; Föllinger, 2010; Lennox and Bolton, 2010; Mourcade, 2008; Johnson, 2005; Lennox, 2001a; Wöhrle, 1999; Kullmann and Föllinger, 1997; Lloyd, 1996; Devereux and Pellegrin, 1990; Gotthelf and Lennox, 1987; Balme, 1987; Gotthelf, 1985; Flashar, 1983, 267–76ff.; 402–11.

3. On Aristotle's works and their survival, see Shields, 2007.

4. Quoted and discussed in Gotthelf, 2012, 346.

5. GA 715a20. On male and female in Aristotle, see Connell, 2016; Mayhew, 2004; Kosman, 2010; Freudenthal, 1995; Kullman, 1999, 118–119.

6. In fact, there are species of hermaphroditic animals (e.g., flatworms and some serranid fishes), as well as a wide range of macroscopic animals that produce parthenogenically at least some of the time. Finally, there are some species of aphid that, so

far as we can tell, lack two sexes entirely, mothers birthing cloned daughters strictly through parthenogenesis. Given the variability of reproduction under differing conditions exhibited in many aphid species, however, a more careful commentator might prefer to say that "no males have yet been observed," which, as we shall see, has the added benefit of being a much more Aristotelian approach.

7. GA 715a23.

8. On spontaneous generation in Aristotle, see Henry, 2016, 2003; Lloyd, 1996, ch. 5; Freudenthal, 1995; Gotthelf, 1989; Lennox, 1982; Balme, 1962.

9. GA at 721a3ff.

10. I have given a fuller treatment of likeness and heredity in Aristotle in Lehoux, 2014, upon which the present account depends. See also Henry, 2006; Coles, 1995; Cooper, 1988.

11. GA 737a8ff. On the interplay of form and matter in generation, see O'Connor, 2015; Byrne, 2015. On the biomechanics of the generation of uniform animal parts, see Lennox, 2014. On the cheese-making analogy, see Ott, 1979.

12. He also allows himself a rare joke at their expense, saying that since a son who resembles his father tends to wear the same kinds of shoes as his father, those who think the seed comes from the whole body speak as though we should suppose that seed comes from the father's shoes as well (GA 723b31).

13. For the ideal case in Aristotle, see GA 766a19; Connell, 2016; Kosman, 2010; Mayhew, 2004; Freudenthal, 1995; Balme, 1987.

14. There is one possible exception (on technical grounds I won't get into here), which is a child's resemblance to an opposite-sex grandparent. See Cooper, 1988; Henry, 2006. See also Lehoux, 2014, where I argue that Aristotle himself seemed to think he had solved the problem pointed to by Cooper and Henry or else had failed to notice it. In any event, he offers us at least one anecdotal example of resemblance to an opposite-sex grandparent at GA 722a10.

15. GA 715b16. Cf. HA 556b22ff., where Aristotle tells us explicitly that lice, fleas, and bedbugs mate and produce nits, but that nothing further is generated by the nits. Cf. also HA 539b10.

16. GA 732b13. Cf. also HA 539b10.

17. The eggs of some common semipelagic Mediterranean fish do grow in size by osmoting water and some minerals from the environment. I thank Tom Johnston at the Ontario Ministry of Natural Resources for clarifying this for me.

18. E.g., at GA 733a2ff.

19. GA 733b14, 758b16, 758b27 et passim. Cf. HA 555b24 et passim.

20. GA 758b7ff., 758b23ff.

21. This last word follows Peck's translation, which avoids the potential confusion in the phrase "third generation."

22. HA 539b9ff.

23. On metamorphosis in Aristotle, see Lloyd, 1996, ch. 5.

24. Another hint that the offspring of spontaneously generated animals remain as larva can be found at GA 723b8.

25. HA 539b13.

26. GA 736a1.

27. GA 736b29ff.

28. Aristotle ascribes different biological functions to different parts of the soul, and not all living things have all parts. Processes of nutrition, growth, and reproduction are the most basic (all living things have these). Sensation and desires are governed by an additional part (all animals have it). Motion seems also to be governed by this part of the soul, although some animals lack the ability to move, so there may be a difference, strictly speaking. Finally comes the one part of the soul that is special to humans, which governs reason and seems, uniquely, to have had some kind of existence separate from matter (see GA 736b28; Metaph. 1070a26).

29. GA 737a18ff.

30. Later thinkers are going to insist that spontaneous generation only occurs in matter that was formerly living, either in the putrefying of now-dead plants and animals or else in the diseased, corrupted flesh of still-living animals and plants. Although Aristotle frequently associates spontaneous generation with putrefaction, he does not explicitly limit either putrefaction or spontaneous generation to formerly living material and in fact allows for the possibility of spontaneous generation in mud (HA 551a4), sand (HA 569a24), dew (HA 551a1), old snow (HA 552b7), and even fire (HA 552b11, although contrast GA 737a1), on which see Macfarlane, 2013.

31. GA 735b1.

32. See Freudenthal, 1995; van der Eijk, 2005; Nutton, 2004; Hankinson, 2003; Berryman, 2002; Long, 1999.

33. GA 735b14ff.

34. A more invasive solution might be to suggest that the sentence has lost τοῦ δ᾽ ἀέρος before the phrase γιγνομένου πνεύματος, meaning "the whitening becomes more compacted as the wateriness within it is separated by heat and the air becomes pneuma," but there is nothing in the *apparatus criticus*, to my eye, that might push quite so far.

35. GA 735b35.

36. At HA 556b28, for example, he says that lice (but not sea lice) are generated from flesh, but he offers no elaboration on the mechanism of this generation in that passage, so its connection to the present context, although suggestive, is not entirely clear. At HA 539a24 he more clearly mentions animals that come to be from residues inside other animals: τὰ δ᾽ ἐν τοῖς ζῴοις αὐτοῖς ἐκ τῶν ἐν τοῖς μορίοις περιττωμάτων, "others [are generated] inside animals themselves, from residues in their organs."

37. Cf. Metaph. 1045a3. Again, though, he offers no theoretical elaboration of the mechanism of such generation (indeed, he does not even say that it is spontaneous in so many words). Part of the reason for his vagueness is that he is calling on the change from corpse to maggots in order to tease out some of the difficulties he has with his language of actuality and potentiality, where he is loath to call a living human "maggots in potential." To get around this, he envisages a kind of middle state mediating between the human and the maggots, where the human is reduced to its matter (εἰς τὴν ὕλην δεῖ ἐπανελθεῖν) "such that if an animal comes from the corpse, it (the living man) goes to matter first, then to an animal," οἷον εἰ ἐκ νεκροῦ ζῷον, εἰς τὴν ὕλην πρῶτον, εἶθ' οὕτω ζῷον.

38. Aristotle distinguishes between two classes of what we would call shellfish: the μαλακόστρακοι (lit., "soft-pottery") and the ὀστρακόδερμοι (lit., "pottery-skinned"), referring to creatures such as lobsters and crayfish, in the first instance, and snails and bivalves, in the second. I will follow common translation practice and parse these as "crustacea" and "testacea," respectively. Testacea for Aristotle include bivalves, shelled gastropods, and sea urchins. See, e.g., HA 527b35ff., GA 763a24ff.

39. HA 569b17.

40. See GA 735b20, 786a7.

41. He also comments on the fact that Aphrodite, the goddess of sexual intercourse, is named after the Greek word for foam, *aphros* (GA 736a20).

42. GA 736a14ff.

43. GA 762b12ff.

44. GA 736b11.

45. See, e.g., De anima 413a21.

46. GA 762a8ff.

47. This refers to the method of propagation of whelks and purpuras, which secrete a slimy liquid that is οἷον ἀπὸ σπερματικῆς φύσεως, "as though from a seed-like nature," but he warns us against seeing it as a true seed (GA 761b31, see also HA 564b19ff.).

48. HA 569b14.

49. HA 569b10.

50. GA 716a13ff.

51. Gotthelf, 1989; Lennox, 1982.

52. GA 762a19ff.

Chapter 2 • Aristotle and Observational Confidence

1. See, e.g., Leroi, 2014; Lloyd, 1996, ch. 5; Lloyd, 1979, 200ff. is particularly good; Kullmann, 1999, 104–7; Kullmann 1974; Flashar, 1983, 403–4, 410–11; Le Blond, 1973; Bourgey, 1955.

2. Lloyd, 1979, ch. 3, makes a good start. The topic is also handled as part of a larger look at Aristotle's methodology in Kullmann, 1999.

3. There are moments when the modern reader is impressed by the care Aristotle takes to clarify or specify a distinction, but it goes without saying that such a response is inevitably grounded in a modern perspective on the ancient text rather than necessarily reflecting the priorities of Aristotle himself.

4. Perhaps the study of chemicals, of rocks, metals, and minerals, or even of medicine might be seen as "sciences of special cases" in a similar way to what I am claiming for Aristotle's biological investigations, but Aristotle does not engage these to anything like the same extent as he does animal species. My point here is to contrast the problems faced by Aristotelian biology with those faced by his approach to physics, for example.

5. Leroi, 2014, for example, thinks that the level of detail in many of the anatomical observations shows that Aristotle himself must have done the dissections on "about 35" different species (59). Still, the level of detail is no guarantee that any given anatomy was Aristotle's own work, even if we do know that he did a great deal of such work generally.

6. PA 666b17–19.

7. GA 764a33–36. This observation is adduced not so much for its own sake as for how it bears on Aristotle's critique of Empedocles and on his own larger theory of sex determination.

8. This and some of Aristotle's other rhetorical qualifications are noted by Le Blond, 1973, 246, albeit briefly.

9. De iuventute et senectute 468b29–30; see also Lloyd, 1979, 216, for commentary and further context.

10. HA 511b13–20.

11. The Polybus passage that he quotes happens to correspond with parts of the Hippocratic texts On the Nature of Man and On the Nature of Bones.

12. HA 513a12–16.

13. On this work and its possible format, see Lennox, 2011; Barnes, 1975, 240; Ross, 1955, 264.

14. HA 496a9. Cf. 508a32. See Kullmann, 1999.

15. Cf. also the "sufficiently observed" proof that wind eggs (by which he means unfertilized eggs, cf. GA 751b22) are not produced from the delayed action of male seed from earlier matings at GA 751a13. They are instead something produced by the female all by herself. See also Kullmann, 1999, 107. Cf. also the description of the chick's development in the Hippocratic On the Nature of the Child, 29.

16. HA 570a3, 570a16–19. Cf. HA 538a4.

17. HA 584b33ff.

18. HA 612b4.

19. GA 757a7. There were a number of different theories on hyena genitalia in antiquity, and Aristotle is here deploying the observation to counter those who believed hyenas to each have two sets of genitalia, male and female.

20. HA 631b3–4.

21. HA 552b7, 559b4.

22. HA 607b21, 605b4.

23. HA 571b33.

24. Libya, HA 606b10; Crete, HA 612a3; Lake Maeotis, HA 620b6, among other examples.

25. HA 569a13ff.

26. HA 569a22ff.

27. Fishermen, HA 547b30; drug-makers, HA 572a22–28; shepherds, GA 767a8; beekeepers, GA 759b20–760a2, etc.

28. GA 756a7.

29. GA 756a32–33. See also Kullmann, 1999, 106ff.; Bourgey, 1955, 89.

30. HA 541a32. Contrast this situation with that of sharks and rays, which take longer to mate: τὰ μὲν οὖν σελάχη πάντα τεθεώρηται ὑπὸ πολλῶν τούτους ποιούμενα τοὺς τρόπους τὴν ὀχείαν· χρονιωτέρα γὰρ ἡ συμπλοκὴ πάντων τῶν ζῳοτόκων ἐστὶν ἢ τῶν ᾠοτόκων, "all the selachia have been observed by many [people] mating in the way described, for the coming together of all vivipara takes longer than that of ovipara" (HA 540b20).

31. The Greek word is ἁλιῆς, which means those who work on the sea. A literal translation would be "saltmen" or perhaps "salts," although the Greek term can include novice sailors in a way that the English *salts* does not. I will use "fishermen" because the majority of Aristotle's uses of the term seem to point to people who work with fish for a living, with the caveat that he may be referring to merchant or military sailors in some instances.

32. HA 532b18–26. See also Kullmann, 1999, 107.

33. HA 535a13–22.

34. Peck, 1970, 73, translates this passage as "scallops, if anyone puts his finger near them while they are open, at once close up," but this does a disservice to the Greek. I suspect that Aristotle's trouble with making sense of the wording came from a simple unawareness that scallops in fact "swim" to escape predators by rapidly clapping their shells open and closed to effect a kind of biological jet propulsion.

35. GA 720b32–35. Interestingly, Aristotle elsewhere seems less certain of how this tentacle is used. See HA 524a5, 541b9, and 544a8, signaling, as most scholars think, a development in his thinking over time rather than a later intrusion into his text by other hands. See also Kullmann, 1999, 107.

36. See Lloyd, 1979, 212; Lloyd 1996, 75.

37. HA 548b11ff.

38. HA 549a4ff.

39. HA 570a3. Cf. HA 538a8.

40. GA 759b20–23. Cf. Pliny, NH XI.46: apium enim coitus visus est numquam, "the mating of bees has never been observed."

41. GA 741a33–37.

42. GA 755b20–22. I take the last clause to involve a reversal of the usual order of reference in a μέν . . . δέ construction (on which, see Denniston, 1954), although this does not solve the main problem with the clause, which is that it is incompatible with HA 538a21, where Aristotle seems to suggest that *channae*, like *erythrinoi*, are only ever full of eggs. There are, however some textual issues with the latter half of the sentence (from αὕτη onward) that may account for the problem. Whether or not textual corruption does explain the surprising invocation of "seminal parts" here, the important thing for our purposes is the statement of fact in the first half of the sentence. On an unrelated note, Peck, in the Loeb HA (2:93 n. a, commenting on 567a27), credits Aristotle with the discovery that some species of sea perch (among which most commentators reckon both the erythrinus and the channa) are hermaphroditic. But neither the passage in question nor any other that I can find in Aristotle calls them hermaphroditic, and so far as I know the fact was not discovered until the late eighteenth century. On the contrary, Aristotle seems to think they are all female. On the hermaphroditism of some species of serranidae, see Smith, 1965. For other passages on the erythrinus in Aristotle (expressing sometimes varying degrees of confidence), see HA 538a20-21, 567a27; GA 741a33–37, 760a8; for its habitat, see HA 598a13.

43. HA 566a27ff.

44. GA 746b5–7. Note that Aristotle alternates between ῥινοβάτης, as here, and ῥινόβατος, as in the previous passage.

45. HA 560a28. For hailstones in cuttlefish eggs, see also HA 525a7.

46. HA 550a20ff.

47. HA 628b14–17.

48. Cf. GA 761a8, where he says another type of wasp has been seen mating "often," in spite of which, as he tells us at HA 629a1, it is unclear whether or not they have stingers.

49. If we compare this passage to others where he uses the singular aorist passive ὤφθη, it is pretty clear that he often uses it to mark singular events: HA 507a22 (singular τι . . . ὤφθη ἔχον τὸν σπλῆνα, even if taken up by a plural τὰ τοιαῦτα, which I take to mean "cases like this" rather than "these cases"), 516a19 (singular ὤφθη ἀνδρὸς κεφαλή), and 565b25 (singular ὤφθη νάρκη μεγάλη).

50. HA 532a16. Cf. PA 683a9.

51. I would like to thank James P. Pitts for his help with this section.

52. HA 542a5ff.

53. E.g., GA 757b23, 760b30, 762a35.

54. See Lloyd, 1987, 141.

55. E.g., HA 559b22, GA 723b19, 750b27, 751a12, 764a34.

56. GA 759a8. For this whole section, cf. HA 553a17ff.

57. Augustine would later hold this theory: *et certe apes semina filiorum non coeundo concipiunt sed tamquam sparsa per terras ore colligunt*, "and of course bees do not get their offspring by mating, but instead collect them by mouth, spread, so to speak, about the countryside" (De trinitate 3.8.13).

58. Because he thinks they have stingers, Aristotle thinks that what we call queen bees are male.

59. GA 759b1.

60. GA 759b25.

61. GA 759b30.

62. Note that this is the closest he ever comes to saying that the *erythrinus* and the *channa* are in any sense hermaphroditic. Because he twice pulls back from saying outright that these fish work exactly the same way as bees, I reject Peck's conclusion that Aristotle anticipated the discovery that sea perch are hermaphroditic (Peck, 1970, commenting on HA 567a27).

63. GA 760a12.

64. GA 759b28, 760a4.

65. GA 760b16.

66. GA 760b27–33.

67. Lloyd, 1979, 137–38.

Chapter 3 • A Blossoming of Creatures

1. See Perkell, 1989; Thomas, 1991; Habinek, 1990; McDonald, 2014; Garani, 2013; Horsfall, 2010; Kronenberg, 2009; Griffin, 2008; Harrison, 2007; Feeney, 2004; Peraki-Kyriakidou, 2003; Gale, 2000. For what is perhaps the most thorough and fascinating survey of the sources, up to and including his contemporaries, see Redi, 1668, 36ff.

2. Varro, De agricultura, 3.16.4. *Pace* Perkell, 1989, 147, I can see no scepticism in Varro's passages on the subject or in Columella (9.14.6–7), who, far from saying that producing bees in this way is impossible, says instead that it is unnecessary (by which I think he means unnecessarily expensive) since bees are prolific enough on their own. Columella complicates the origins of the idea by attributing it to "Democritus" (often, in Columella, meaning the obscure Bolos of Mendes) and to Mago, the now-lost Carthaginian writer on agriculture.

3. Vergil, Georgics, 4.295–316.

4. I am switching some of the verbs from passive to active in this passage. To my ear the passives make the English sound stilted where the Latin is anything but.

5. Della Porta, 1589, 2.2.

6. Of course, *vilior* need not be translated as "cheaper" here; we could also say "more common," although I think that the two effectively amount to the same thing in the end.

7. See, e.g., Perrone Compagni, 2007.

8. Kruk, 1990, 275.

9. Hasse, 2006, 2007; Bertolacci, 2013a, 2013b.

10. On spontaneous generation in the Arabic tradition, see Bertolacci, 2013a; Zambelli, 2008; Freudenthal, 2002; Hasse, 2007, 2006; Kruk, 1990; Genequand, 1984. On the Giver of Forms, see, e.g., Hasse, 2012.

11. For the medieval Latin discussion, see van der Lugt, 2004, 170ff.; Hasse, 2006. For early modern discussion of human spontaneous generation—a question that tended to focus on the origins of the newly discovered Native Americans, and a fascinating topic that must unfortunately be passed over here in the interests of space, see Gliozzi, 2000.

12. On the Lucretian account of spontaneous generation, see Campbell, 2003. On Lucretian atomism generally, see Beretta, 2015; Gatzemeier, 2013; Lehoux, Morrison, and Sharrock, 2013; Beretta and Citti, 2008; Gale, 2007; Gillespie, 2007; Fowler, 2002; Volk, 2002; Sedley, 1998. On the reception of Lucretius in the Renaissance and early modern period, see Norbrook, Harrison, and Hardie, 2016; Palmer, 2014; Brown, 2010; Passannante, 2011; Gillespie, 2007.

13. For technical reasons, and partly to avoid Stoic-like predetermination, the Epicureans also posited an occasional random "swerve" that could affect atoms and cause them to change direction spontaneously. (On the meaning of *spontaneous* in this context, see Johnson, 2013.)

14. Lucretius, DRN 5.805–8.

15. See Sedley, 1998; Garani, 2013.

16. Lucretius, DRN 5.837–41.

17. Lucretius, DRN 5.812–13.

18. Lucretius, DRN 5.797–800.

19. Lucretius, DRN 5.793. Cf. 2.1154–55.

20. Lucretius, DRN 5.826–27.

21. This is particularly significant because issues around the final cause are going to dog Aristotelian theories of spontaneous generation in late antiquity and the Middle Ages. On teleology in Aristotle's biology, see, e.g., Gotthelf, 2012; Leunissen, 2010; Johnson, 2005; Lennox, 2001a; Gotthelf and Lennox, 1987. On the Epicurean rejection of teleology, see, e.g., Sedley, 2007.

22. An Aristotelian would certainly read it this way, and Platonists and Stoics, who also have accounts of causation that are ultimately teleological (see, e.g., Sedley, 2010, 2007; Frede, 2003; Hankinson, 1999), would likely take *finis* in its technical philosophical sense as well and so might be caught off guard by Lucretius in exactly the same way.

23. Lucretius, DRN 2.887–88.

24. Lucretius, DRN 2.897–901.

25. He uses words that mean "uniting as a whole" (*concilio*) as well as "yoking together" (*coniungo*, which we have already seen) for the relationship between souls and bodies throughout the De rerum natura.

26. The Latin word *effetus* (from *ex-fetus*) specifically means "worn out from having had too many babies." I can think of no English equivalent.

27. Lucretius, DRN 2.872, 3.719–20, 2.1150–51.

28. GA 762a21.

29. GA 736b30.

30. On Augustine, see, e.g., Vessey 2012; O'Donnell, 2006; Fitzgerald, 1999. For Augustine's account of spontaneous generation, see van der Lugt, 2004.

31. On Augustine's reading of creation, see Knuuttila, 2014.

32. Augustine, De Gen. ad litt. 5.6.19.

33. His most thorough treatment of these *rationes seminales* is in the De Genesi ad litteram, but he also invokes them elsewhere in his corpus. On Augustinian *rationes seminales,* see Hirai, 2008, 2005; van der Lugt, 2004; Colish, 1985; Agaësse and Solignac, 1972, 1:659ff.; Meyer, 1914.

34. Augustine, De Gen. ad litt. 5.7.20.

35. This is particularly interesting because *primordia* had been Lucretius's standard word for *atoms*, although Augustine cannot mean the same thing by it.

36. Augustine, De Gen ad litt. 3.14.22.

37. Augustine, De Gen. ad litt. 3.14.23

38. Augustine, De Gen. ad litt. 6.10.17.

39. Augustine, De Gen. ad litt. 5.7.20.

40. Augustine, De trinitate 3.9.16.

41. On my use of *protogenesis* to cover what is often called preexistence by historians and philosophers of biology, see ch. 6.

42. Augustine, De Gen ad litt. 3.14.22. The Latin material in square brackets is simply resupplying grammatical subject and direct object explicitly mentioned in an earlier sentence and assumed as given by Augustine here.

43. καθάπερ . . . θεούς = Heraclitus fr. A9 DK.

44. PA 645a15–26. See Kullmann, 1999, 104–5; Lloyd, 1987, 1979.

45. Lit., "that for the sake of which."

46. Augustine, De trinitate 3.8.13.

47. Augustine, De trinitate 3.8.13.

Chapter 4 • *Inheritance and Innovation*

1. See Metaph. 7.7–9. For discussion, see Menn, sec. 2g2.

2. E.g., De caelo 289a20ff.

3. See Henry, 2016, 2003; Meyrav, 2016; Wilberding, 2012; Hasse, 2007; Brague, 1999; Bertolacci, 2013a.

4. Meyrav, 2016.

5. Themistius, In metaph. 12. Translation following Henry, 2016, with one modification (wasps for hornets [which are in any case a type of wasp] following the near-universal agreement of the tradition). I apologize for not including the Hebrew, which in Moses Ibn Tibbon's 1255 translation is the earliest extant version of the complete text. I unfortunately do not have the language, or enough Arabic yet to work from Averroes's slightly earlier quotation of this same passage.

6. Phys. 194b13.

7. Averroes, Metaph. 1464–65, Genequand, 1984, trans.

8. Averroes, Metaph. 1502, Genequand, 1984, trans.

9. See, e.g., Metaph. 1032b13.

10. See, e.g., Ptolemy, Tetr. 1.4, 1.8, 1.9.

11. Indeed, it would be another century or two after Aristotle before this kind of astrological explanation would take root in the Greek-speaking world. See Barton, 1994.

12. Averroes, De substantia orbis, Hyman, 1986, trans. (modified slightly), 95.

13. We know that he did not adopt every aspect of Greek astrology, refusing to accept, for instance, that there was any such thing as a maleficent astral influence. See Freudenthal, 2009; Saliba, 1999; Sezgin, 1979; Ullmann, 1972.

14. To be sure, Aristotle does enable the sun, by means of its varying degrees of heat throughout the year, to alter the amount of moisture found at the surface of the earth, but this is still considerably more restricted than what Averroes is claiming. See Meteor. 340b19ff. Aristotle also allows the moon to somehow control the periodicity and timing of some biological functions in the GA, but he does not give it a role in causing generation specifically. See GA 777b17ff. and GA 767a1ff.

15. Genequand, 1984, 30.

16. This refrain is also shorthand for another point Aristotle brings up repeatedly: that in natural things like animals the formal and efficient causes of generation are the same, being the actual parent.

17. Elsewhere in his corpus, Aristotle uses the phrase ἄνθρωπος ἄνθρωπον γεννᾷ or a variant wording at least ten times, and this is the only instance where καὶ ὁ ἥλιος gets appended. On the importance of the phrase ἄνθρωπος ἄνθρωπον γεννᾷ for Aristotle's accounts of causation and nature, see Oehler, 1969.

18. On these, see Menn, secs. 2g2 and esp. 3b2a, and I thank Menn for alerting me to this connection. See also Henry, 2015; Sedley 2010.

19. See Möhle, 2015; di Giovanni, 2014; Bertolacci, 2013b; Twetten and Baldner, 2013; Weijers, 2005; Honnefelder et al., 2005; Weisheipl, 1980.

20. Albertus, De min., 2.3.2.

21. A version of the argument goes back at least as far as Cicero, who imagined bringing an Antikythera-type planetarium to "savage" Britain—even the inhabitants of that wild land would recognize it as the deliberate product of a rational mind (Cicero, DND 2.83).

22. *Parisiis* following Wyckoff, 1967, and Albertus's own De caus. prop. elem. 2.2.5. The text has *parvis.*

23. Albertus, De min. 1.1.2.

24. Galen, On Antecedent Causes, 7.82.

25. Albertus, De min. 2.3.1. See Wyckoff, 1967, 128 n. 2.

26. Albertus, De min. 2.3.1.

27. Albertus, De min. 2.3.3.

28. HA 538a1ff.

29. Given the emphasis on the location of the hairs and worms, the ἔχω must be taken in this more specific sense, since finding them without a womb would mean the eels still "have them," but they are not offspring.

30. Albertus, De animal. 24.8.

31. E.g., De animal. 2.77: *piscis qui Graece vocatur enchelyz, quod est anguilla. sunt autem et aput nos plurimi tales*, "the fish that the Greeks call enchelyz, that is, the eel; there are many of these in our land"; De animal. 24.8: *in aliis autem omnibus aquis Germaniae anguillae multae inveniuntur*, "In all other German waters many eels are found."

32. Albertus, De animal. 1.61.

33. Albertus, De animal. 6.80.

34. Kitchell and Resnick, 1999, 563.

35. Albertus, De animal. 6.71.

36. Albertus, De animal. 6.82.

37. He has also been noted as "in all likelihood the most prolific author of the whole of the Middle Ages" (Kitchell and Resnick, 1999, 18; van Steenberghen, 1932, 529). On his "nearly illegible" handwriting, see Kitchell and Resnick, 1999, 34; Ostlender, 1952.

38. GA 741a33ff.

39. Albertus, De animal. 16.103.

40. For my fixation on garlic and magnets, see Lehoux, 2012.

41. Della Porta, Mag. nat. 2.4.

42. Della Porta, Mag. nat. 2.1, 3.1, and 2.2, respectively.

43. Della Porta, Mag. nat. 3.1.

44. Della Porta, Mag. nat. 3.1.

45. Della Porta, Mag. nat. 3.1.

46. Contrast, though, Aristotle's claim at GA 763b22 that "in the most perfect of animals, male and female are separate," where he seems to have in mind something like what will, after Aquinas, become the most common usage.

47. See Hasse, 2006.
48. Albertus, De animal. 21.51.
49. Albertus, De animal. 13.3.
50. Albertus, De animal. 13.23, 13.26. Cf. 13.33, 13.74.
51. Albertus, De animal. 13.119.
52. Albertus, De animal. 14.9.
53. Albertus, De animal. 21.5, 21.36.
54. Albertus, De animal. 15.43.
55. Albertus, De animal. 14.3–4.
56. Cf. also Albertus, De animal. 16.8.
57. Albertus, De animal. 16.5.
58. Aquinas, Metaph., 1400–1401.
59. Phys. 2.4–6; Metaph. 1032a28.
60. Aquinas, Metaph. 1403; translation following Rowan, 1961, in some places.
61. On Buridan's position, see Hasse, 2006.
62. On Liceti, see Hirai, 2011, 2006; Blank, 2010; Castellani, 1968, Ongaro, 1964.
63. Liceti, 1618, 2.1.
64. Indeed, I cannot make much sense of what Aristotle does say is going on in this instance except that it is clearly a case of sexual generation for him, and he thinks of it as "the greatest proof," μέγιστον δὲ τεκμήριον, that the semen does not come from the parts of the body at GA 723b19—again, though, he is unclear how he thinks this proof works.
65. Aristotle mentions the propagation of plant cuttings at GA 723b16, but I cannot find him discussing the case of insects anywhere, apart from a brief mention in the De anima (411b18ff.), where he says that plants can be cut up and continue living, and "certain insects" when divided will have each half showing sensation and movement. However, he qualifies this latter statement to add that this happens "only for a time," ἐπί τινα χρόνον. This is quite different from Liceti's claim that this is a form of "generation," by which *multa viventia constituantur*, "many [new] living things come to be" (Liceti 1618, 2.1), which sounds more like something that could only have been seriously held of animals after Trembely's experiments on polyps (see ch. 6, below).
66. GA 736b28–29.
67. He returns to a basic twofold division later in the chapter.
68. Liceti, 1618, 2.36.
69. Liceti, 1618, 2.37.
70. It also drives my translation of *porro* in the passage just above, reading it as an emphatic adverb (soul is in the material at a [metaphysical] distance) rather

than as a simple conjunction ("furthermore" or "besides, such a soul in that material as in a container"). I freely admit, though, that the latter reading would be perfectly plausible.

71. On the overlap between spontaneous generation theories and the formation of metals and stones, see, e.g., Hirai, 2005.

72. Liceti, 1618, 2.41.

73. Liceti, 1618, 2.41.

74. Liceti, 1618, 1.11.

Chapter 5 • Interlude: Is Life Special?

1. I include this possibility in order to find a space for Aristotle's theory, on which, see below.

2. On Descartes' biological mechanism, see, e.g., Sanhueza, 1997; Jedidi, 1991; Bitbol-Hespériès, 1990; Gaukroger, 2000; Manning, 2015; Aucante, 2006; Des Chene, 2001; Duchesneau, 1998; Grmek, 1990; Farley, 1974.

3. Bastian, 1870, 171.

4. PA 681a10ff. Cf. also HA 531b9, where Aristotle says that the sea anemone is also like a plant. In Albertus's paraphrase of the PA passage, at De animal. 13.119, he adds that the intermediate forms are *animalia, sed non perfecte et vere*, "animals, but not perfectly and truly [so]."

5. HA 588b4ff.

6. ἔμψυχον literally means "ensouled," but it usually implies liveliness and animation beyond simply having the property of life. In the current passage, where plants appear unliving when compared to animals, Aristotle is clearly aiming at drawing out a stronger emphasis in this set of distinctions, and so I think "lively" is as fair a translation as any.

7. GA 762a19.

8. *Mixture* is an important technical term in ancient physics and alchemy, and there are several theories about what it means for two substances to be "truly mixed." On Aristotle's theory of mixtures, see Viano, 2015; Cooper, 2004, ch. 7; Wood and Weisberg, 2004. On pneuma see also Freudenthal, 1995.

9. GA 736a1.

Chapter 6 • Toward a Showdown

1. Cudworth, 1678, 148. Emphasis his here and throughout.

2. He builds the idea on Jacob Schegk's earlier *facultas plastica*. On plastic natures, see Hirai, 2011, 2007; Giglioni, 1994.

3. Cudworth, 1678, 150.

4. Cudworth, 1678, 147. The Galen passage he mentions is at De usu partium XVII, K361–2.

5. On microscopy in this period, see Ratcliff, 2009; Generali and Ratcliff, 2007; Schickore, 2007; Ruestow, 1996; Fournier, 1996; Wilson, 1995.

6. See, e.g., Parke, 2014.

7. Redi, 1668, 27–28: *puzzolenti, infracidate, e corrotte . . . tutti convertiti in un acqua grossa, e torbida . . . [e] l' anguile . . . rigonfiando e ribollendo ed a poco a poco perdendo la figura.*

8. Redi, 1668, 142.

9. Redi, 1668, 143–44.

10. I noted this in an earlier chapter, but in case the reader missed it, Redi's long discussion of bees and cattle is fascinating (see Redi, 1668, 36ff.) and well worth reading.

11. Redi, 1668, 138.

12. Redi, 1668, 140–41.

13. Redi, 1668, 145.

14. Harris, 1704, quoted in Mendelsohn, 1976.

15. Chambers, 1728, entry under "equivocal generation."

16. Paulo da Venezia, Lectura super librum De anima, ad 2, comm. 47 (quoted in van der Lugt, 2004, 133 n. 101, with *musce* for my *muscae*).

17. On Gassendi's theory of generation, see Fisher, 2006, 2004; Hirai, 2005, 2003; Duchesneau, 1998.

18. Gassendi, De gen. animal. 4.1 (Opera 2, 263a).

19. Gassendi calls them *animulae* rather than *animae* for a reason.

20. Sometimes it was created as just a soul, sometimes as a homunculus or as body parts in some arrangement or other. I am bulldozing a little bit here, as there is a range of nuanced positions that different historical actors took about what, exactly, had been formed before conception, but given the focus of the present book I hope the reader will forgive these necessary generalizations.

21. Harvey, 1651, 2. The Empedocles fragment is B79 DK and is quoted by Aristotle at GA 731a9.

22. Following Willis, 1847.

23. Harvey, 1651, 82–83.

24. Journal des sçavans, 6 Mar. 1679, 65.

25. This terminology follows Roger, 1963; Farley, 1974; and Bowler, 1971, who distinguish epigenesis from preformation and preexistence theories. Some authors do not consistently follow Roger's definitions, which can generate confusion if the reader is not careful (Fisher, 2006, for example, deliberately uses *preformation* more broadly).

26. Bowler, 1971, gives a detailed history of these last three theories.

27. On Needham's work on spontaneous generation, see Stefani, 2002; Mazzolini and Roe, 1986; Roe, 1983; Mazzolini, 1976.
28. Needham, 1748, 636.
29. Needham, 1748, 637–38.
30. Needham, 1748, 639.
31. Needham, 1748, 639.
32. Needham, 1748, 643.
33. Needham, 1748, 643.
34. Needham, 1748, 657.
35. Needham, 1748, 623.
36. Needham, 1748, 624.
37. Needham, 1748, 629, emphasis in original.
38. Needham, 1748, 627, 632.
39. Needham, 1748, 630.
40. Needham, 1748, 631.
41. Gould, 1829, leads the reader through this experiment as follows: "A curious experiment may be performed by separating one of the larger sort [of paste eel] from the rest by placing it in another drop of water by means of a fine point of a quill; it may then be easily cut asunder by a fine lancet and if the division is made about the middle of the animal, several oval bodies will be seen to come forth . . . These are the young, curled up in a fine membrane: the largest and most forward break through it, unfold themselves, and swim away" (25).
42. Needham, 1748, 632.
43. Needham, 1748, 659–60.
44. Needham, 1748, 656.
45. Needham, 1748, 652.
46. Needham, 1748, 656–57. Cf. Needham, 1750, 243ff.
47. Needham, 1748, 645.
48. Needham, 1748, 657.
49. Needham, 1748, 653.
50. Needham, 1748, 654.
51. Needham, 1748, 658–59.
52. On Needham's metaphysics, see Mazzolini and Roe, 1986.
53. Spallanzani, 1765, 69–70.
54. Spallanzani, 1765, 78.
55. Needham, 1748, 637.
56. Spallanzani, 1765, 80.
57. Spallanzani, 1765, 83.
58. Spallanzani, 1765, 84.
59. Spallanzani, 1765, 83–84.

60. Following Needham's French translation, *ses molécules élastiques*. Spallanzani, 1769, 132.

61. Spallanzani, 1765, 81.

62. Needham, 1769, 139.

63. Needham, 1748, 621–22.

64. Needham, 1769, 207–8. The oft-quoted Horace passage is from *Ars* 191.

65. Needham, 1769, 211–12.

66. Needham, 1769, 212.

67. Hacking, 1983, informs much of my analysis throughout this section.

68. Spallanzani, 1765, 78.

69. Needham, 1769, 216–17.

70. We do not believe that life can come into existence in the same way as Needham did, of course, but we do agree that life can come from nonlife (and indeed has done so at least once), and we also reject protogenesis. Precisely how to parse out the fertilization of an animal egg or the fission of a bacterium in terms of epigenesis versus (something like) preformation is beyond the scope of this volume, but a case can certainly be made for epigenesis. See Maienschein, 2005.

Conclusion

1. The literature on scientific realism is vast, and I can only touch on most of it here. Consult Lehoux, 2012 (esp. chs. 9 and 10), for how I see the historical record as complicating (but not endangering) a belief in scientific realism.

2. Hacking, 2000; Stanford, 2006, 2013.

3. See the papers in Soler, Trizio, and Pickering, 2015, for the current state of the field.

4. Duhem, 1906.

5. If we now object that we have a better set of criteria for what counts as predictively successful, that is all well and good. What we cannot know is whether our criteria are finally sufficient to warrant confidence. They thought theirs were sufficient; we now think they were wrong and that ours are sufficient instead. As always, both of those judgments are historically contingent. Both are subject to change or rejection. Both are inevitably forward-blind.

Agaësse, P., and A. Solignac, trans. and comm. (1972) Augustin: La Genèse au sens littéral en douze livres, Paris.

Algra, K., J. Barnes, J. Mansfeld, and M. Schofield, eds. (1999) The Cambridge History of Hellenistic Philosophy, Cambridge.

Aucante, V. (2006) "Descartes's Experimental Method and the Generation of Animals," in J. E. H. Smith, 65–79.

Balme, D. M. (1987) "The Place of Biology in Aristotle's Philosophy," in Gotthelf and Lennox, 9–20.

Balme, D. M. (1962) "Development of Biology in Aristotle and Theophrastus: Theory of Spontaneous Generation," Phronesis 7, 91–104.

Barnes, J. (1975) Aristotle's Posterior Analytics, Oxford.

Barton, T. (1994) Ancient Astrology, New York.

Bastian, H. C. (1870) "Facts and Reasonings Concerning the Heterogenous Evolution of Living Things," Nature 2, 170–77, 219–28.

Beretta, M. (2015) La rivoluzione culturale di Lucrezo, Rome.

Beretta, M., and F. Citti, eds. (2008) Lucrezio, la natura e la scienza, Florence.

Bernardi, W. (1986) Le metafisiche dell' embrione, Florence.

Berryman, S. (2002) "Aristotle on Pneuma and Animal Self-Motion," Oxford Studies in Ancient Philosophy 23, 85–97.

Bertolacci, A. (2013a) "Averroes against Avicenna on Human Spontaneous Generation," in A. Akasoy and G. Giglioni, eds., Renaissance Averroism and Its Aftermath, Dordrecht, 37–54.

Bertolacci, A. (2013b) "Avicenna's and Averroes's Interpretations and their Influence on Albertus Magnus," in Resnick, 95–136.

Bigelow, M., trans. (1909) Experiments on the Generation of Insects, Francesco Redi of Arezzo, Chicago.

Bigwood, J. M. (1993) "Aristotle and the Elephant, Again," American Journal of Philology 114, 537–55.

Bitbol-Hespériès, A. (1990) Le principe de vie chez Descartes, Paris.

Blank, A. (2010) "Material Souls and Imagination in Late Aristotelian Embryology," Annals of Science 67, 187–204.
Blank, A. (2007) "Composite, Substance, Common Notions, and Kenelm Digby's Theory of Animal Generation," Science in Context 20, 1–20.
Bourgey, L. (1955) Observation et expérience chez Aristote, Paris.
Bowler, P. (1971) "Preformation and Pre-existence in the Seventeenth Century," Journal of the History of Biology 4, 221–44.
Brague, R. (1999) Thémistius: Paraphrase de la Métaphysique d'Aristote, Paris.
Brams, J. (2003) La riscoperta di Aristotele in Occidente, Milan.
Brown, A. (2010) The Return of Lucretius to Renaissance Florence, Cambridge, MA.
Byrne, C. (2015) "Compositional and Functional Matter," Apeiron 48, 387–406.
Campbell, G. L. (2003) Lucretius on Creation and Evolution, Oxford.
Castellani, C. (1968) "Le problème de la *generatio spontanea* dans l'oeuvre de Fortunio Liceti," Revue de synthese 89, 323–40.
Chambers, E. (1728) Cyclopaedia, London.
Coady, C. A. J. (1992) Testimony: A Philosophical Study, Oxford.
Cohen, H. F. (2010) How Modern Science Came into the World, Amsterdam.
Coles, A. (1995) "Biomedical Models of Reproduction in the Fifth Century BC and Aristotle's 'Generation of Animals,'" Phronesis 40, 48–88.
Colish, M. L. (1985) The Stoic Tradition from Antiquity to the Middle Ages, Leiden.
Connell, S. M. (2016) Aristotle on Female Animals, Cambridge.
Cooper, J. M. (2004) Knowledge, Nature, and the Good, Princeton, NJ.
Cooper, J. M. (1988) "Metaphysics in Aristotle's Embryology," Proceedings of the Cambridge Philological Society 214, 14–41.
Cudworth, R. (1678) The True Intellectual System of the Universe, London.
Cunningham, A. (2000) "The Identity of Natural Philosophy," Early Science and Medicine 5, 259–78.
Cunningham, A. (1991) "How the Principia Got Its Name," History of Science 29, 377–92.
Cunningham, A. (1988) "Getting the Game Right," Studies in the History and Philosophy of Science 19, 365–89.
Cunningham, A., and P. Williams (1993) "De-centring the 'Big Picture,'" British Journal for the History of Science 26, 407–32.
Deichmann, U. (2012) "Origin of Life," History and Philosophy of the Life Sciences 34, 341–60.
De Leemans, P. (2010) "Aristotle Transmitted: Reflections on the Transmission of Aristotelian Scientific Thought in the Middle Ages," International Journal of the Classical Tradition 17, 325–53.
Denniston, J. D. (1954) The Greek Particles, 2nd ed., Oxford.

Des Chene, D. (2001) Spirits and Clocks, Ithaca, NY.
Des Chene, D. (2000) Life's Form, Ithaca, NY.
Detlefsen, K. (2006) "Explanation and Demonstration in the Haller-Wolff Debate," in J. E. H. Smith, 235–61.
Devereux, D., and P. Pellegrin, eds. (1990) Biologie, logique et métaphysique chez Aristote, Paris.
di Giovanni, M. (2014) "The Commentator: Averroes' Reading of the Metaphysics," in F. Amerini and G. Galluzzo, eds., A Companion to the Latin Medieval Commentaries on Aristotle's Metaphysics, Leiden, 59–94.
Duchesneau, F. (2006) "'Essential Force' and 'Formative Force': Models for Epigenesis in the 18th Century," in B. Feltz, M. Crommelinck, and P. Goujon, eds., Self-Organization and Emergence in Life Sciences, Berlin, 171–86.
Duchesneau, F. (1998) Les modèles du vivant de Descartes à Leibniz, Paris.
Duhem, P. (1906) La théorie physique: Son objet et sa structure, Paris.
Ebrey, D., ed. (2015) Theory and Practice in Aristotle's Natural Science, Cambridge.
Farley, J. (1974) The Spontaneous Generation Debate from Descartes to Oparin, Baltimore.
Feeney, D. (2004) "Interpreting Sacrificial Ritual in Roman Poetry," in A. Barchiesi, J. Rüpke, and S. Stephens, eds., Rituals in Ink, Stuttgart, 1–21.
Fisher, S. (2006) "The Soul as Vehicle for Genetic Information," in J. E. H. Smith, 103–23.
Fisher, S. (2004) "Gassendi's Atomist Account of Generation and Heredity in Plants and Animals," Perspectives on Science 11, 484–512.
Fitzgerald, A., ed. (1999) Augustine through the Ages, Grand Rapids, MI.
Flashar, H. (1983) "Aristoteles," in H. Flashar, ed., Die Philosophie der Antike, vol. 3, Ältere Akademie, Aristoteles-Peripatos, Basel, 175–458.
Föllinger, S., ed. (2010) Was ist "Leben?" Aristoteles' Anschauungen zur Entstehung und Funktionsweise von Leben, Stuttgart.
Fournier, M. (1996) The Fabric of Life: Microscopy in the Seventeenth Century, Baltimore.
Fowler, D. (2002) Lucretius on Atomic Motion, Oxford.
Frede, D. (2003) "Stoic Determinism," in Inwood, 179–205.
French, R., and A. Cunningham. (1996) Before Science, Aldershot, UK.
Freudenthal, G. (2009) "The Astrologization of the Aristotelian Cosmos," in A. C. Bowen and C. Wildberg, eds., New Perspectives on Aristotle's De caelo, Leiden, 238–81.
Freudenthal, G. (2002) "The Medieval Astrologization of Aristotle's Biology," Arabic Sciences and Philosophy 12, 111–37.
Freudenthal, G. (1995) Aristotle's Theory of Material Substance, Oxford.
Gale, M. (2000) Virgil on the Nature of Things, Cambridge.

Gale, M., ed. (2007) Lucretius, Oxford.

Garani, M. (2013) "Empedoclean Cows and Sheep," in Lehoux, Morrison, and Sharrock, 233–60.

Gasking, E. (1967) Investigations into Generation, 1651–1828, Baltimore.

Gatzemeier, S. (2013) Ut ait Lucretius: Die Lukrezrezeption in der lateinischen Prosa bis Laktanz, Göttingen.

Gaukroger, S. (2000) "The Resources of a Mechanist Physiology and the Problem of Goal-Directed Processes," in J. Schuster, S. Gaukroger, and J. Sutton, eds., Descartes' Natural Philosophy, New York, 383–400.

Genequand, C. (1984) Ibn Rushd's Metaphysics, Leiden.

Generali, D., and M. Ratcliff, eds. (2007) From Makers to Users: Microscopes, Markets, and Scientific Practices in the Seventeenth and Eighteenth Centuries, Florence.

Giglioni, G. (1994) "*Spiritus plasticus* between Pneumatology and Embryology," Studia comeniana et historica 24, 83–90.

Gillespie, C. (2007) The Cambridge Companion to Lucretius, Cambridge.

Gliozzi, G. (2000) Adam et le Nouveau Monde, Paris.

Gotthelf, A. (2012) Teleology, First Principles, and Scientific Method in Aristotle's Biology, Oxford.

Gotthelf, A. (1989) "Teleology and Spontaneous Generation in Aristotle," Apeiron 22, 181–93.

Gotthelf, A., ed. (1985) Aristotle on Nature and Living Things, Pittsburgh.

Gotthelf, A., and J. G. Lennox, eds. (1987) Philosophical Issues in Aristotle's Biology, Cambridge.

Gould, C. (1829) Companion to the Microscope, with Full Directions for Preparing the Vegetable Infusions to Produce Animalcules, 5th ed., London.

Goy, I. (2014) "Epigenetic Theories," in I. Goy and E. Watkins, eds., Kant's Theory of Biology, Berlin, 43–60.

Grattan-Guinness, I. (2004) "The Mathematics of the Past: Distinguishing Its History from Our Heritage," Historia Mathematica 31, 163–85.

Griffin, J. (2008) "The Fourth Georgic, Virgil and Rome," in Volk, 2008, 225–48.

Grmek, M. D. (1990) La première révolution biologique, Paris.

Habinek, T. (1990) "Sacrifice, Society, and Vergil's Ox-Born Bees," in M. Griffith and D. Mastronarde, eds., Cabinet of the Muses, Atlanta, 209–24.

Hacking, I. (2000) "How Inevitable Are the Results of Successful Science?" Philosophy of Science 67, S58–S71.

Hacking, I. (1983) Representing and Intervening, Cambridge.

Hankinson, R. J. (2003) "Stoicism and Medicine," in Inwood, 295–309.

Hankinson, R. J. (1999) "Explanation and Causation," in Algra et al., 476–512.

Harris, J. (1704) Lexicon technicum, London.

Harrison, S. J. (2007) Generic Enrichment in Vergil and Horace, Oxford.
Harvey, W. (1651) Exercitationes de generatione animalium, London.
Hasse, D. N. (2012) "Avicenna's 'Giver of Forms' in Latin Philosophy, Especially in the Works of Albertus Magnus," in D. N. Hasse and A. Bertolacci, eds., The Arabic, Hebrew, and Latin Reception of Avicenna's Metaphysics, Berlin, 225–50.
Hasse, D. N. (2007) "Spontaneous Generation and the Ontology of Forms in Greek, Arabic, and Medieval Latin Sources," in P. Adamson, ed., Classical Arabic Philosophy, London, 150–75.
Hasse, D. N. (2006) Urzeugung und Weltbild, Hildesheim.
Henry, D. (2016) "Themistius and the Problem of Spontaneous Generation," in Sorabji, 179–94.
Henry, D. (2015) "Aristotle on the Cosmological Significance of Biological Generation," in Ebrey, 100–118.
Henry, D. (2006) "Aristotle on the Mechanism of Inheritance," Journal of the History of Biology 39, 425–55.
Henry, D. (2003) "Themistius and Spontaneous Generation in Aristotle," Oxford Studies in Ancient Philosophy 24, 183–207.
Hirai, H. (2011) Medical Humanism and Natural Philosophy, Leiden.
Hirai, H. (2008) "Logoi Spermatikoi and the Concept of Seeds in the Mineralogy and Cosmogony of Paracelsus," Revue d'histoire des sciences 61, 245–64.
Hirai, H. (2007) "La main de Dieu entre l'âme at la nature," Journal de la Renaissance 5, 205–22.
Hirai, H. (2006) "Âme de la terre, génération spontanée et origine de la vie," Bruniana & Campanelliana 12, 451–69.
Hirai, H. (2005) Le concept de semence dans les théories de la matière à la renaissance, Turnhout, BE.
Hirai, H. (2003) "Le concept de semence de Pierre Gassendi entre les théories de la matière et les sciences de la vie au XVIIe siècle," Medicina nei scuoli 15, 205–26.
Honnefelder, L., R. Wood, M. Dreyer, and M.-A. Aris, eds. (2005) Albertus Magnus und die Anfänge der Aristoteles-Rezeption im lateinischen Mittelalter, Münster.
Horsfall, N. (2010) "Bees in Elysium," Vergilius 55, 39–45.
Howlett, P., and M. S. Morgan, eds. (2010) How Well Do Facts Travel? Cambridge.
Hyman, A. (1986) Averroes' De substantia orbis, Cambridge, MA.
Inwood, B., ed. (2003) The Cambridge Companion to the Stoics, Cambridge.
Jedidi, A. (1991) Les fondements de la biologie cartésienne, Paris.
Johnson, M. R. (2013) "Nature, Spontaneity, and Voluntary Action in Lucretius," in Lehoux, Morrison, and Sharrock, 99–130.

Johnson, M. R. (2005) Aristotle on Teleology, Oxford.

Kitchell, K. F., and I. M. Resnick (1999) Albertus Magnus: On Animals, Baltimore.

Knuuttila, S. (2014) "Time and Creation in Augustine," in D. V. Meconi and E. Stump, eds., The Cambridge Companion to Augustine, Cambridge, 81–97.

Kosman, A. (2010) "Male and Female in Aristotle's Biology," in Lennox and Bolton, 147–67.

Kronenberg, L. (2009) Allegories of Farming from Greece and Rome, Cambridge.

Kruk, R. (1990) "A Frothy Bubble: Spontaneous Generation in the Medieval Islamic Tradition," Journal of Semitic Studies 35, 265–82.

Kuhn, T. S. (1962) "The Structure of Scientific Revolutions," International Encyclopedia of Unified Science: Foundations of the Unity of Science, vol. 2.2, Chicago.

Kullmann, W. (1999) "Aristoteles' wissentschaftliche Methode in seinen zoologischen Schriften," in Wöhrle, 103–23.

Kullmann, W. (1974) Wissenschaft und Methode, Berlin.

Kullmann, W., and S. Föllinger, eds. (1997) Aristotelische Biologie, Stuttgart.

Kusch, M. (2002) Knowledge by Agreement, Oxford.

Le Blond, J. M. (1973) Logique et méthode chez Aristote, Paris.

Lehoux, D. (2014) "Why Doesn't My Baby Look Like Me? Likeness and Likelihood in Ancient Theories of Reproduction," in V. Wohl, ed., Probabilities, Hypotheticals, and Counterfactuals in Ancient Greek Thought, Cambridge, 208–29.

Lehoux, D. (2012) What Did the Romans Know? Chicago.

Lehoux, D., A. D. Morrison, and A. Sharrock, eds. (2013) Lucretius: Poetry, Philosophy, Science, Oxford.

Lennox, J. G. (2014) "Aristotle on the Emergence of Material Complexity," HOPOS 4, 272–305.

Lennox, J. G. (2011) "Aristotle's Biology," in the Stanford Encyclopedia of Philosophy.

Lennox, J. G. (2006) "The Comparative Study of Animal Development," in J. E. H. Smith, 21–46.

Lennox, J. G. (2001a) Aristotle's Philosophy of Biology, Cambridge.

Lennox, J. G. (2001b) Aristotle: On the Parts of Animals, Cambridge.

Lennox, J. G. (1982) "Teleology, Chance, and Aristotle's Theory of Spontaneous Generation," Journal of the History of Philosophy 20, 219–38.

Lennox, J. G., and R. Bolton, eds. (2010) Being, Nature, and Life in Aristotle, Cambridge.

Lennox, J. G., and K. Kampourakis (2013) "Biological Teleology," in K. Kampourakis, ed., The Philosophy of Biology, Dordrecht, 421–54.

Leroi, A. M. (2014) The Lagoon: How Aristotle Invented Science, New York.

Leunissen, M. (2010) Explanation and Teleology in Aristotle's Science of Nature, Cambridge.

Liceti, F. (1618) De spontaneo viventium ortu libri quatuor, Padua.

Lloyd, G. E. R. (1996) Aristotelian Explorations, Cambridge.

Lloyd, G. E. R. (1987) The Revolutions of Wisdom, Berkeley.

Lloyd, G. E. R. (1979) Magic, Reason, and Experience, Cambridge.

Lloyd, G. E. R. (1970) Early Greek Science, New York.

Long, A. A. (1999) "Stoic Psychology," in Algra et al., 560–84.

Macfarlane, P. (2013) "Aristotle on Fire Animals," Apeiron 46, 136–65.

Maienschein, J. (2005) "Epigenesis and Preformation," in the Stanford Encyclopedia of Philosophy.

Manning, G. (2015) "Descartes' Metaphysical Biology," HOPOS 5, 209–39.

Mayhew, R. (2004) The Female in Aristotle's Biology, Chicago.

Mazzolini, R. (1976) "Two Letters on Epigenesis from John Turberville Needham to Albrecht von Haller," Journal of the History of Medicine and the Allied Sciences 31, 68–77.

Mazzolini, R., and S. Roe. (1986) Science against the Unbelievers, Oxford.

McDonald, J. A. (2014) Orpheus and the Cow (diss., Cornell University), Ithaca, NY.

Mendelsohn, E. I. (1976) "Philosophical Biology vs Experimental Biology," in M. Greene and E. I. Mendelsohn, eds., Topics in the Philosophy of Biology, 37–65.

Menn, S. (n.d.) The Aim and the Argument of Aristotle's Metaphysics, available at https://www.philosophie.hu-berlin.de/de/lehrbereiche/antike/mitarbeiter/menn.

Meyer, H. (1914) Geschichte der Lehre von den Keimkräften, Bonn.

Meyrav, Y. (2016) "Spontaneous Generation and Its Metaphysics in Themistius' Paraphrase of Aristotle's Metaphysics 12," in Sorabji, 195–210.

Mocek, R. (1995) "Caspar Friedrich Wolffs Epigenesis-Konzept," Biologisches Zentralblatt 114, 179–90.

Möhle, H. (2015) Albertus Magnus, Münster.

Mouracade, J. (2008) Aristotle on Life, Kelowna, BC.

Needham, J. T. (1769) Notes de M. Needham, in Spallanzani, 1769, 139–235.

Needham, J. T. (1750) Nouvelles observations microscopiques, Paris.

Needham, J. T. (1748) "A Summary of Some Late Observations upon the Generation, Composition, and Decomposition of Animal and Vegetable Substances," Philosophical Transactions of the Royal Society 490, 615–66.

Norbrook, D., S. Harrison, and P. Hardie, eds. (2016) Lucretius and the Early Modern, Oxford.

Nutton, V. (2004) Ancient Medicine, New York.

O'Connor, S. (2015) "The Subjects of Natural Generations in Aristotle's Physics I.7," Apeiron 48, 45–75.

O'Donnell, J. (2006) Augustine: A New Biography, New York.
Oehler, K. (1969) "Ein Mensch zeugt einen Menschen," in K. Oehler, Antike Philosophie und byzantinisches Mittelalter, Munich, 95–145.
Ongaro, G. (1964) "La gernerazione e il 'moto' del sangue nel pensiero di F. Liceti," Castalia 20, 75–94.
Ostlender, H. (1952) "Die Autographe Alberts des Grossen," in H. Ostlender, ed., Studia Albertina, Münster, 3–21.
Ott, S. (1979) "Aristotle among the Basques: The 'Cheese Analogy' of Conception," Man 14, 699–711.
Palmer, A. L. G. (2014) Reading Lucretius in the Renaissance, Cambridge, MA.
Parke, E. C. (2014) "Flies from Meat and Wasps from Trees: Reevaluating Francesco Redi's Spontaneous Generation Experiments," Studies in the History and Philosophy of Biological and Biomedical Sciences 45, 34–42.
Passannante, G. P. (2011) The Lucretian Renaissance, Chicago.
Peck, A. L., trans. (1970) Aristotle: History of Animals, Books 4–6, Cambridge, MA.
Peraki-Kyriakidou, H. (2003) "The Bull and the Bees," Les études classiques 71, 151–74.
Perkell, C. (1989) The Poet's Truth, Berkeley.
Perrone Compagni, V. (2007) "Un ipotesi non impossibile," Bruniana & Campanelliana 13, 99–111.
Ratcliff, M. J. (2009) The Quest for the Invisible, Aldershot, UK.
Redi, F. (1668) Esperienze intorno alla generazione degl' insetti, Florence.
Resnick, I. M., ed. (2013) A Companion to Albert the Great, Leiden.
Roe, S. (1983) "John Turberville Needham and the Generation of Living Organisms," Isis 74, 159–84.
Roe, S. (1981) Matter, Life, and Generation, Cambridge.
Roe, S. (1979) "Rationalism and Embryology," Journal of the History of Biology 12, 1–43.
Roe, S. (1975) "The Development of Albrecht von Haller's Views on Embriology," Journal of the History of Biology 8, 167–90.
Roger, Jaques. (1963) Les sciences de la vie dans la pensée française du XVIIIe siècle, Paris.
Ross, W. D. (1955) Parva naturalia, Oxford.
Rowan, J. P., trans. (1961) Aquinas: Commentary on the Metaphysics of Aristotle, Chicago.
Ruestow, E. G. (1996) The Microscope in the Dutch Republic, Cambridge.
Saliba, G. (1999) "Astronomy and Astrology in Medieval Arabic Thought," in R. Rashed and J. Biard, eds., Les doctrines de la science de l'antiquité à l'âge classique, Louvain, 131–64.
Sanhueza, G. (1997) La pensée biologique de Descartes dans ses rapports avec la philosophie scholastique, Paris.

Schickore, J. (2007) The Microscope and the Eye, Chicago.
Sedley, D. (2010) "Teleology, Aristotelian and Platonic," in Lennox and Bolton, 5–29.
Sedley, D. (2007) Creationism and Its Critics in Antiquity, Berkeley.
Sedley, D. (1998) Lucretius and the Transformation of Greek Wisdom, Cambridge.
Sezgin, F. (1979) Geschichte des arabischen Schrifttums, vol. 7, Astrologie, Meteorologie und Verwandtes, Leiden.
Shapin, S. (1996) The Scientific Revolution, Chicago.
Shields, C. (2007) Aristotle, New York.
Smith, C. L. (1965) "The Patterns of Sexuality and the Classification of Serranid Fishes," American Museum Novitates 2207 (29 Jan.), 1–20.
Smith, J. E. H., ed. (2006) The Problem of Animal Generation in Early Modern Philosophy, Cambridge.
Soler, L., E. Trizio, and A. Pickering. (2015) Science As It Could Have Been, Pittsburgh.
Sorabji, R., ed. (2016) Aristotle Re-Interpreted: New Findings on Seven Hundred Years of the Ancient Commentators, London.
Spallanzani, L. (1769) Nouvelles recherches sur les découvertes microscopiques et la génération des corps organisés, London.
Spallanzani, L. (1765) Saggio di osservazioni microscopiche concernenti il sistema della generazione de signori di Needham e Buffon, Modena.
Stanford, P. K. (2013) "Underdetermination of Scientific Theory," in the Stanford Encyclopedia of Philosophy.
Stanford, P. K. (2006) Exceeding Our Grasp, Oxford.
Stanford Encyclopedia of Philosophy, Stanford, CA, https://plato.stanford.edu/.
Stefani, M. (2002) Corruzione e generazione: John T. Needham e l' origine del vivente, Florence.
Thomas, R. (1991) "The 'Sacrifice' at the End of the Georgics, Aristaeus, and Vergillian Closure," Classical Philology 86, 211–18.
Thompson, D. W. (1913) On Aristotle as a Biologist, Oxford.
Tipton, J. A. (2014) Philosophical Biology in Aristotle's Parts of Animals, New York.
Twetten, D., and S. Baldner. (2013) "Introduction to Albert's Philosophical Work," in Resnick, 165–72.
Ullmann, M. (1972) Die Natur- und Geheimwissenschaften im Islam, Leiden.
van der Eijk, P. (2005) Medicine and Philosophy in Classical Antiquity, Cambridge.
van der Lugt, M. (2004) Le ver, le démon et la vierge, Paris.
van Steenberghen, F. (1932) "Saint Albert le Grand, docteur de l'église," Collectanea mechliniensia 21, 518–34.
Vessey, M. (2012) A Companion to Augustine, Oxford.

Viano, C. (2015) "Mixis and Diagnôsis: Aristotle and the 'Chemistry' of the Sublunary World," Ambix 62, 203–14.

Volk, K. (2002) The Poetics of Latin Didactic, Oxford.

Volk, K., ed. (2008) Vergil's Georgics, Oxford.

von Lippmann, E. O. (1933) Urzeugung und Lebenskraft, Berlin.

Weijers, O. (2005) "The Literary Forms of the Reception of Aristotle," in Honnefelder et al., 555–84.

Weisheipl, J. A. (1980) "The Life and Works of St. Albert the Great," in J. A. Weisheipl, ed., Albertus Magnus and the Sciences, Toronto, 13–51.

Wilberding, J. (2012) "Neoplatonists on Spontaneous Generation," in J. Wilberding and C. Horn, eds., Neoplatonism and the Philosophy of Nature, Oxford, 196–213.

Willis, R., trans. (1847) The Works of William Harvey, M.D., London.

Wilson, C. (1995) The Invisible World, Princeton, NJ.

Wöhrle, G., ed. (1999) Biologie (Geschichte der Mathematik und der Naturwissenschaften in der Antike, vol. 1), Stuttgart.

Wood, R., and M. Weisberg. (2004) "Interpreting Aristotle on Mixture," Studies in the History and Philosophy of Science 35, 681–706.

Wyckoff, D. (1967) Albertus Magnus: Book of Minerals, Oxford.

Zambelli, P. (2008) "Sono gli autoctoni generati 'per accidens' o 'a casu'? Note sulla generazione spontanea dell'uomo," Giornale critico della filosofia italiana 87, 30–58.

INDEX